BRITISH SCIENTISTS AND
THE MANHATTAN PROJECT

W0051416

Also by Ferenc Morton Szasz

THE DIVIDED MIND OF PROTESTANT AMERICA

RELIGION IN THE WEST (*editor*)

THE DAY THE SUN ROSE TWICE: The Story of the Trinity Site Nuclear Explosion, July 16, 1945

THE PROTESTANT CLERGY IN THE GREAT PLAINS AND MOUNTAIN WEST

British Scientists and the Manhattan Project

The Los Alamos Years

Ferenc Morton Szasz
Professor of History
University of New Mexico, Albuquerque

MACMILLAN

© Ferenc Morton Szasz 1992

Softcover reprint of the hardcover 1st edition 1992

All rights reserved. No reproduction, copy or transmission of
this publication may be made without written permission.

No paragraph of this publication may be reproduced, copied or
transmitted save with written permission or in accordance with
the provisions of the Copyright, Designs and Patents Act 1988,
or under the terms of any licence permitting limited copying
issued by the Copyright Licensing Agency, 33–4 Alfred Place,
London WC1E 7DP.

Any person who does any unauthorised act in relation to this
publication may be liable to criminal prosecution and civil
claims for damages.

First published 1992 by
MACMILLAN ACADEMIC AND PROFESSIONAL LTD.
Houndmills, Basingstoke, Hampshire RG21 2XS
and London
Companies and representatives
throughout the world

ISBN 978-1-349-12733-7   ISBN 978-1-349-12731-3 (eBook)
DOI 10.1007/978-1-349-12731-3

A catalogue record for this book is available
from the British Library.

For Margaret and Maria

Contents

Preface

This is a study of the British scientific mission to Los Alamos, New Mexico, from 1943 to 1947, and the impact it had on the early history of the atomic age. In retrospect, it is clear that this book had a dual origin. During the late 1970s and early 1980s, I researched and wrote an account of the world's first atomic explosion (July 16, 1945) at Trinity Site, on the Alamogordo Air Base in central New Mexico. In 1984, the University of New Mexico Press published this as *The Day the Sun Rose Twice*. The research for this project, especially the opportunity to interview many of the original participants of the Manhattan Project, lured me forever into the fascinating world of atomic history.

During the academic year of 1985–6, I taught at the University of Exeter in Devon as a Visiting Fulbright Professor for the American and Commonwealth Arts Division. There my family and I became fascinated with British life, especially that of the West Country. This study of the British scientists who went to wartime Los Alamos is an attempt to combine these two themes.

Any project that stretches over a decade acquires numerous obligations. Consequently, I would like to begin by thanking the following: Don Pierce, Gerald Lane, and Charles Bates, three Albuquerque book dealers who always kept their eyes open for anything that touched on atomic history; Brenda Spurlock Landstedt, who proved that an undergraduate work-study person could do graduate level research; Dennis Bilger of the Harry S Truman Library in Independence, Missouri, for gracious treatment during my visit there; the staff at the Franklin D. Roosevelt Library in Hyde Park, New York, for making available the "Atomic Bomb folder"; Edward J. Reese of the Modern Military Branch of the National Archives, Washington, DC; the staff at the Public Record Office in Kew, London; Anna Broussard and Dorothy Wonsmos of Zimmerman Library, the University of New Mexico; Mrs Ashley C. Oates for generously serving as a "broker" to put me in touch with her parents, Sir Ernest Titterton and Peggy Titterton; Sarah (Polly) Tuck for sharing memories of her parents, James and Elsie Tuck; Los Alamos veteran Robert Porton for loaning me his priceless collection of the Laboratory daily bulletins for 1943–5; Mrs Lorna Arnold of the United Kingdom Atomic Energy Authority for responding so splendidly to a stranger's phone call; and Professor W.N. Everitt of the Department of Mathematics, University of Birmingham, for sending me materials on his late colleague, Tony Skyrme. Reference librarian Kate

Luger deserves special commendation for devising increasingly intricate ways of tracking down missing British scientists. Thanks to her, I was eventually able to account for every member of the British Mission who came to Los Alamos from 1943 to 1947.

Research in the early history of the atomic era is ever entwined with current matters of state. On both sides of the Atlantic, national security measures are such that certain World War II materials still remain closed to historians. This is especially true for the British side of the story, where numerous documents are unlikely to be available within the foreseeable future. On the American side, materials that had been declassified in the 1950s have suddenly been reclassified again. (The rise of international terrorism prompted this move. While the nuclear technology of the World War II years may be outdated, it still works.) Consequently, a special word of thanks must go to Roger A. Meade, head Archivist/Historian at the Los Alamos National Laboratories. Meade did yeoman work in uncovering virtually everything in the archives that pertained to the British experience in New Mexico. In addition, he and his staff provided invaluable assistance in transcribing the many interviews that I conducted over the years.

I would also like to credit my colleague, Dr Donald Skabelund, and Paul W. Henricksen of Los Alamos for their aid; the late Dr Ralph Carlisle Smith, who read an earlier draft; and Professors A.P. French and Janet Roebuck, who read through the entire manuscript. After almost thirty years, I am still indebted to Richard W. Smith, Professor Emeritus at Ohio Wesleyan University, for his inspiration and guidance. Richard G. Robbins, Barton J. Bernstein and Robert Del Tredici were all most generous in sharing primary documents. Two anonymous reviewers for St. Martin's Press saved me from a number of potentially embarrassing errors. The typing skills of Pat E. Devejian and Penelope P. Katson were at their usual level of excellence. So, too, were Margaret Connell Szasz's highly valued editorial skills.

Special mention should also be made to our colleagues and friends at the University of Exeter: Mick and Nancy Gidley, Richard Maltby, David and Gillian Horn, Ron and Anne Tamplin, and Peter and Louisa Quartermain. They patiently answered all our questions about British life, even the foolish ones, without a hint of embarrassment.

I could not have completed this study without the courtesy of those who graciously allowed me to interview them regarding their knowledge of the British Mission to Los Alamos: Ms Ellen Bradbury, Dr Norris Bradbury, Mrs Hanni Bretscher, Ms Judy Florence, Dr Anthony P. French, Dr D.J. Littler, Dr Donald C. Marshall, Dr J. Carson and Kathleen Mark, Dr John Manley, Dr P.B. Moon, Sir Rudolf Peierls, Lord William George Penney, Dr M.J. Poole, Mrs Mollie Rodriguez, Dr Joseph Rotblat, Dr Alice Kimball

Smith, Dr Cyril Stanley Smith, Dr Ralph Carlisle Smith, Sir Ernest and Peggy Titterton. Of course, all scholarship treating these issues must begin with the works by Margaret Gowing, the official historian of the British atomic world: Margaret Gowing, *Britain and Atomic Energy*, 1939–1945 (London: Macmillan, 1964); Margaret Gowing, *Independence and Deterrence*, Vol. I, *Policy Making*; Vol II, *Policy Execution* (London: Macmillan, 1974); and her American counterparts, who wrote the official history of the Atomic Energy Commission: Richard G. Hewlett and Oscar E. Anderson, Jr., *The New World, 1939/1946*: Vol. I of *A History of the United States Atomic Energy Commission* (University Park, Penn.: Pennsylvania State University Press, 1962); Richard G. Hewlett and Francis Duncan, *Atomic Shield, 1947/1952*: Vol. II of *A History of the United States Atomic Energy Commission* (University Park, Penn., and London: Pennsylvania State University Press, 1969); Richard G. Hewlett and Jack M. Holl, *Atoms for Peace and War, 1953–1961: Eisenhower and the Atomic Energy Commission*: Volume III of *A History of the Atomic Energy Commission* (Berkeley: University of California Press, 1989).

Finally, I would like to thank all the members of my family. My father, Ferenc P. Szasz, Ing. (1906–1984), introduced me to both the world of science and that of émigré life. My mother, Mary Szasz, received updates on this project almost every week for the last five years. Chris, Scott and Eric also heard about it with regularity. And I would especially like to thank Margaret and Maria for listening so patiently to the endless stories about the British Mission to Los Alamos. In an academic household, the gift of listening is far more precious than can ever be imagined.

FMS

Introduction

The first weeks of August, 1945, proved the most decisive in the history of humankind. Or so thought Isaiah Berlin, then head of the Special Survey Section of the British Embassy in Washington, DC. Berlin would later go on to become an internationally famous philosopher/historian, but in 1945 he was drafting manuscript dispatches for the British home front. On August 11, he wrote of America's astonishment at a week of "earth shattering events": Potsdam, Hiroshima, the Soviet declaration of war on Japan, Nagasaki, and President Harry S Truman's radio address to the nation disclosing the secret of the atomic bomb. These revelations, Berlin noted, had combined to make the United States "more breathless than jubilant." A weapon culled from the pages of Buck Rogers and Flash Gordon had entered into history. The morality of using it on innocent civilians, plus the immediate necessity for arms control, had both become widespread subjects of popular discussion.

Berlin's dispatches from Washington also noted that the success of the atomic bomb was widely recognized as a joint American–British–Canadian enterprise. President Harry S Truman praised the British efforts, and comments from both British scientists and former Prime Minister Winston Churchill were widely publicized. As one American scientist observed, "There is no need to talk of an Anglo-American federation – it is here."[1] On the day of Hiroshima, a prominent Washington journalist wrote Truman's private secretary to suggest that because the atomic weapon had been a joint American–British–Canadian enterprise, it should henceforth be termed the "ABC" bomb. Truman seems to have toyed with the idea.[2]

This, however, was not to be. Within a few weeks, the "A" in "A bomb" came to symbolize not only "atomic" but also "American." Film-makers, journalists and writers soon forgot about the British and Canadian contributions to the greatest industrial and scientific enterprise of the twentieth century.

On August 9, 1945, the day of Nagasaki, President Truman approved the release of the official, semi-technical account of the building of the bomb, the "Smyth Report" (so named for its author, Princeton physics professor Henry D. Smyth). Written without consultation with either the British or Canadians, the Smyth Report circulated as a litho-printed document for about a month. Then, on September 10, Princeton University Press rushed an inexpensive cloth-bound version into print. To everyone's surprise, sales topped 100,000 within a year. It was not

long before the Smyth Report was translated into every major language –
including a Russian printing of 30,000 copies. This book provided the
world with its first account of the most incredible scientific discovery of the
modern age. Officially entitled *Atomic Energy for Military Purposes*, the
subtitle reflected its emphasis: "The Official Report on the Development of
the Atomic Bomb under the Auspices of the United States Government,
1940–1945."[3]

Simultaneously, the government of Great Britain issued its own version
of this momentous scientific breakthrough, entitled *Statements Relating to
the Atomic Bomb*. A pamphlet of forty pages, this booklet described the
British role in the quest for atomic weapons.[4] The pamphlet contained a brief
statement by current Prime Minister Clement Attlee, a longer, previously
prepared comment by former Prime Minister Winston Churchill, and a
concise summary of the major events of atomic history. This last section
was written in a twenty-four hour period by Tube Alloys secretary Michael
Perrin. Perrin closed his twenty-page account with the movement of the
British atomic scientists to join the American Manhattan Project in late
1943. British Physicist Philip B. Moon was so eager to distribute *Statements
Relating to the Atomic Bomb* to his American colleagues at Los Alamos that
he offered to cover personally the expense of 500 copies if they were sent
to him there.[5]

The British forty-page pamphlet, however, could never compete with
Smyth's 260-page book. For several years the Smyth Report remained the
primary popular account of the making of the atomic bomb.

Several British scientists expressed dismay when they first read their
copies of the Smyth Report. They argued that Smyth had slighted their role
in the atomic enterprise. Eventually the British converted the editors of the
Princeton University Press to their cause. Starting with the November 1945
printing, the press added *Statements Relating to the Atomic Bomb* (plus a
brief comment by the Canadian government) as an "Appendix" to the
Smyth Report.[6] This forty-page addendum, however, did little to alter the
general impression that the development of the atomic bomb had been
primarily an American enterprise.

British pique was not easily mollified. Immediately after the war, Imperial
Chemical Industries (ICI) engineer W.A. Akers, an early leader of the
United Kingdom nuclear effort, observed that because the British accomplishments could not be calculated in terms of money, it was impossible to
make them appear as large as the Americans'.[7] Two years later, a reviewer
in the British journal *Nature* complained that MGM's hastily shot film version
of the atomic story, "The Beginning or the End," had left the erroneous
impression that almost all the atomic scientists had been American.[8]

Not until fifteen years later did British historians move to correct this imbalance. In 1961, science writer Ronald W. Clark subtitled his popular study, *The Birth of the Bomb*, "The Untold Story of Britain's Part in the Weapon That Changed the World."[9] Three years later, Margaret Gowing published the first volume of her definitive official history, *Britain and Atomic Energy, 1939–1945*. Gowing's account received rave reviews in the British press, but it reached a largely academic audience. Her study had little impact on the popular mind, especially in the United States. When Professor Gowing visited the Norris Bradbury Science Museum in Los Alamos in 1983, she expressed considerable dismay. References to the British contribution to the Project were either misleading or completely ignored.[10]

Over the years, American scholars have continued along the path blazed by the Smyth Report. James Phinney Baxter III's *Scientists Against Time*, which won the 1947 Pulitzer prize in history, praised the British scientists for their accomplishments in the creation of radar and other technical items. But Baxter concentrated almost exclusively on the American role in the development of the bomb itself.[11]

In 1949, former Manhattan Project head, General Leslie R. Groves, wrote a four-page letter devoted to the British role in the Manhattan Project. Arguing against the proposed plan to share nuclear data with them, Groves downplayed their impact. "I cannot recall any direct [British] contribution to our success in achieving the bomb," he said. When he wrote his memoirs in 1962, *Now It Can Be Told*, Groves termed the British contribution "helpful but not vital."[12]

Perhaps the most extreme statement appeared in the thirty-five-volume *Official History of the Manhattan Project* (written in 1946/7 but not widely available until 1977). This study suggested that the British scientists involved were of "moderate attainments" and that they crossed the Atlantic primarily to obtain nuclear information from the Americans. The British effort, the anonymous author concluded:

> . . . was in no sense vital and actually not even important. To evaluate it quantitatively at one percent of the total would be to overestimate it. The technical and engineering contribution was practically nil. Certainly it is true that without any contribution at all from the British, the date of our final success need not have been delayed by a single day.[13]

This is absurd. The British contributions to the development of the atomic bomb were crucial. Had the members of the British Mission at Los Alamos not done what they did when they did it, the atomic bomb would not have been available to end the war in August of 1945. An atomic weapon surely would have been discovered later – 1947, perhaps, or 1955, or whenever. But the world would have been vastly different in any case.

As numerous historians have reminded us, the battles of World War II were not waged exclusively on the fields of combat. World War II also drew forth a clash of scientific minds. In their respective laboratories, the Allied scientists raced their German counterparts for a wide variety of technical secrets. World War II was "A Wizard War" with "Scientists Against Time" involved in "a race on the edge of time."[14] As Margaret Gowing has noted, scientific discoveries cannot be hidden forever; but they can be *delayed*. In many cases – certainly World War II was one of them – the time sequence proved of utmost importance.[15] British Mission wife Peggy Titterton phrased the issue for her generation: "I'm glad Hitler didn't get it first."[16]

The program to build an atomic weapon eventually grew into, perhaps, the largest organized effort since the construction of the ancient pyramids. It employed over 200,000 workers and cost over two billion dollars. The giants of American industry, DuPont, Kellex, Stone and Webster, plus the nation's major universities, joined both American-born and émigré scientists in "one of the greatest military-industrial adventures in history."[17]

Within this massive operation, however, a historian may single out three factors that were absolutely critical to its success: the leadership of J. Robert Oppenheimer, director of Los Alamos; the supervision of General Leslie R. Groves, overall head of the Project; and the contributions of the British Mission to Los Alamos.

The story of Oppenheimer is widely known. A child prodigy, Oppenheimer graduated from Harvard *magna cum laude* in three years instead of four, and then embarked on four years of study in Europe. He earned his PhD from Göttingen in 1927. In 1929, Oppenheimer assumed a joint appointment in physics at the University of California in Berkeley and the California Institute of Technology in Pasadena, where he dazzled students in both schools with his wit and brilliance. His colleague, I.I. Rabi, often declared that Oppenheimer had the best mind in physics for his generation. Although he seemed an unusual choice as director of Los Alamos – he had strong left-wing political connections and no prior administrative experience – he soon proved his detractors wrong. Not only did he keep abreast of all the scientific progress of the lab, his sensitivity to the needs of his fellow workers led him to numerous acts of personal kindness. "We just worshipped him," Peggy Titterton recalled. "We would have done anything for that man."[18]

Even Oppenheimer's fall from grace in 1954, when he lost his security clearance after the famous Oppenheimer hearings, has not tarnished the accomplishments of his Los Alamos years. Oppenheimer's haggard appearance and characteristic "porkpie" hat rose to almost mythic proportions in postwar America. When one thinks of the Manhattan Project, one invariably thinks of Oppenheimer first. "A lesser man couldn't have done it," remarked Norris Bradbury, his successor as Los Alamos Lab director.[19]

The name of Leslie R. Groves is almost as widely known as that of Oppenheimer's. Son of a Presbyterian Army chaplain, Groves ran the sprawling Manhattan Project as his personal fiefdom. Unlike Oppenheimer, Groves proved rather difficult to get along with. While he never raised his voice or used foul language, he criticized constantly and often ran roughshod over his subordinates. In addition, he lacked any trace of a sense of humor. Groves' brusque and heavy handed manner, combined with his rotund shape and aloofness, made him a convenient target for disparaging comment. Even those who criticized him, however, respected his organizing skills. And those who worked closely with him held him in the highest esteem. Los Alamos patent lawyer Ralph Carlisle Smith marvelled at how Groves could cut through any red tape to get things done; Kenneth D. Nichols, Groves' chief assistant, labelled him an excellent administrator; and his private secretary, Jane O'Leary, absolutely adored him.[20]

Groves' role in the Manhattan Project can hardly be overestimated. More than any other person, Groves provided the needed focus for all 1942–45 nuclear-related activity. This was crucial because the potential for atomic power and/or atomic weapons was born from the splitting of the same atom. In 1942, the nuclear world contained a variety of still unproven possibilities: individual power sources for ships, tanks or airplanes; large-scale power plants for urban areas; a fission weapon; radio isotopes for medical experiments; a radioactive "poison" to halt the invasion of enemy armies (an issue that was seriously discussed in the planning for D-Day); unlimited fusion (hydrogen) power; or the possibility of a fusion (hydrogen) weapon.

Groves realized that all these goals could not be accomplished simultaneously. Thus, he established a set of priorities that instituted a clear hierarchy. While the Army would consider other issues as they arose, everything had to be subordinated to the production of an atomic bomb. As Australian physicist Marcus L. Oliphant remembered, "It was emphasized again and again that the prime object was the production of a military weapon in the shortest possible time."[21]

Groves' reputation also plummeted after the war. For seventeen months after V-J day, he presided over the nation's atomic matters as a lame duck commander. His conservative political views, which were of minor importance during the war itself, began to loom much larger afterwards.

In 1946, he banished all French scientists from the Canadian nuclear installations because he felt them too left-wing. From 1945 to 1948, he insisted
that the American military should remain the sole custodian of the nation's
nuclear arsenal. In private he spoke disparagingly about the British. In
1954, at the height of the Cold War, he told Congress that "Russia was our
enemy and ... the [Manhattan] project was conducted on that basis."[22]
When he wrote his autobiography in 1962, he did so to counter the ambiguity he sensed in the popular understanding of the history of the atomic
bomb.[23]

Groves' ego proved almost as annoying to postwar nuclear administra-
tors as his politics. The general not only took credit for all past events, he
also gave advice on all present events, too. In early 1946, when the "first
team" of Los Alamos scientists was preparing to leave for their regular,
peace-time university assignments, Groves rushed to Los Alamos to persuade
them to remain. Loudspeakers were set up for the address, which everyone
assumed would be a call to stay for patriotic reasons. But, as engineer Ray
Powell recalled, "what we heard instead was a monologue of how great
General Groves was; as a result the exodus from Los Alamos accelerated."[24]

Once the Atomic Energy Commission (AEC) took command of the
nation's nuclear operations in January of 1947, they made no secret of
wanting Groves out.[25] In a later reminiscence, Harry Truman recalled the
postwar Groves as "a pain in the neck" (surely a euphemism).[26] In 1948,
Groves finally retired from the Army to go to work for Remington Rand
Corporation. He died in 1970. Although Leslie R. Groves never achieved
the popular reputation of the other military figures of his era, Dwight
Eisenhower, Douglas MacArthur, George Patton, or Viscount Montgomery,
he has fared somewhat better with historians. Without Groves, all agree, the
Manhattan Project would never have succeeded.[27]

The story of the British Scientific Mission to Los Alamos forms the third
"indispensable" factor in the complex and tragic tale of the building of the
first atomic bombs. But it is considerably harder to define. These twenty
scientists have never had the high profile or controversial lives of either
Oppenheimer or Groves. The foremost members of the British team –
Rudolf Peierls and Otto Frisch – were basically shy men. The most notorious, Soviet spy Klaus Fuchs, was shyer still. The most visible figure,
Danish Nobel Laureate Niels Bohr, failed in his quest to internationalize the
atom. Another prominent British scientist, Joseph Rotblat, voiced opposition to nuclear weapons and nuclear power for years. Consequently, the
saga of the British Mission to Los Alamos has proven difficult to highlight.

The impact of the British Mission lay primarily in three areas, all linked
by continuity of personnel. First, the initial research and reports of the

1939–1941 period (the unjustly neglected Otto Frisch-Rudolf Peierls Memorandum, plus the Maud Reports) convinced American scientists that an atomic weapon could be constructed before the probable end of hostilities. Second, after the British and American projects were merged in 1943, the contributions of the twenty scientists stationed at Los Alamos (including both Frisch and Peierls, plus James Chadwick and Philip Moon from the Maud Committee) insured that the bomb was available "in the shortest possible time." Without either contribution, World War II would not have ended as early as August of 1945.

But the story does not stop there. Not only did the members of the British Mission determine the date that hostilities ended, they also shaped the contours of the immediate postwar world. There is a direct British Mission link to the emergence of the first three nuclear powers: the United States in July and August, 1945; the Soviet Union in August, 1949; and the United Kingdom in October, 1952. In addition to this technical legacy, the men from Los Alamos moved slowly into politics. There they helped create a still wider legacy: the establishment of the Pugwash Conferences in 1955; a growing debate over nuclear power; and a deeper understanding of the dilemmas of the "nuclear culture" they had done so much to create. Thus, the score of scientists whom the British sent to Los Alamos helped forge the nuclear boundaries of the mid-twentieth century. Their story is worth telling.

Depending on how one counts, the British Mission to Los Alamos included approximately two dozen scientists. In alphabetical order, they were: Egon Bretscher, James Chadwick, Anthony P. French, Otto R. Frisch, Klaus Fuchs, James Hughes, Derrik J. Littler, William G. Marley, Donald G. Marshall, Philip B. Moon, Rudolf E. Peierls, William G. Penney, George Placzek, Michael J. Poole, Joseph Rotblat, Harold Sheard, Tony H.R. Skyrme, Ernest W. Titterton, and James L. Tuck.[28]

In addition, several other scientists also fell under the umbrella of the British Mission. Although officially under the auspices of the Canadian government, J. Carson Mark's salary was paid by the British, and Carson and Kathleen Mark were generally considered as part of the British Mission. The most famous member was Danish physicist Niels Bohr, who arrived with his son Aage. The Bohrs never lived permanently on "the Hill" as Los Alamos was called, but, beginning in December, 1943 they made several extended visits to the lab as "consultants." One might also mention Sir Geoffrey Taylor, an expert in the theory of blast waves, who visited Los Alamos in May of 1944; and Lord Cherwell (Frederick Lindemann), Winston Churchill's chief scientific adviser, who also visited the lab for a day in October, 1944.

The average person is not likely to recognize these names. This situation may be fairly summarized by an incident that occurred while I was researching this study in the Zimmerman Library of the University of New Mexico. When a reference specialist asked me what I was working on, I replied: "The British Mission to wartime Los Alamos." She seemed puzzled. "I never realized that the Church of England was so concerned about its scientists," she said.[29] The following story is an attempt to clarify this situation.

1 Background

The British connection with Los Alamos had a rather tenuous beginning. In July of 1938, Austrian-born Lise Meitner, the foremost woman physicist of her generation, fled Germany because of Hitler's racial laws. After a short stay in Holland, she moved to Stockholm, where she found employment at the Nobel Institute for Physics. Just before Christmas, Meitner received a rather puzzling letter from her longtime German collaborator, physical chemist Otto Hahn. Working with his colleague Fritz Strassmann, Hahn had bombarded uranium with slow neutrons. Their experiment seemed to produce unexpected traces of barium in the result (barium is about half the atomic weight of uranium). Conventional wisdom suggested that a neutron bombardment of uranium would, at best, "chip off" fragments of material that would lie very close to uranium in the high end of the periodic table. Trained as physical chemists, Hahn and Strassmann felt slightly unsure of themselves in physics, and thus sought out Meitner's expertise in the matter.[1]

Shortly afterwards, Meitner's nephew, Otto R. Frisch, who had fled Germany in 1933, made plans to visit her for the Christmas vacation. They met at Kungalv, a small town north of Göteborg on the west coast of Sweden, and Meitner first brought the matter up at breakfast. Later in the day, they continued the discussion as they took a brief hike through the snow. Meitner first had to steer the conversation away from Frisch's latest interest – a plan to build a large magnet – but she soon involved him in the problem. Drawing on Niels Bohr's "liquid drop" model of the nucleus, Meitner and Frisch came to an astonishing conclusion: the uranium nucleus in the Hahn-Strassmann experiment had captured a neutron, elongated into sort of a Victorian "wasp waist," and finally split into two more or less equal parts. Because of their mutual repulsion, these two nuclei then flew apart with great force. Unbeknown to themselves, Hahn and Strassmann had split the uranium atom in two.[2] Excited, Meitner and Frisch laid plans to send their explanation to the prestigious British scientific magazine, *Nature*.

As soon as Frisch returned to his work at Niels Bohr's Institute in Copenhagen, he sought out a visiting American biologist, William Arnold, to inquire about the proper English term for the division of cells. Thus, the word "fission" appeared in a totally new context when *Nature* printed their "letter to the editor" on February 11, 1939.[3] At first the new use of the noun "fission" seemed awkward, for there was no accompanying verb form. But

1

the February issue of *Nature* began a process that would eventually make "fission" into a household word.[4] Meitner chose to remain in Sweden for the duration of the war, but both Frisch and Bohr became bulwarks of the British Mission to Los Alamos.

When Frisch explained his theory to Bohr, the latter exclaimed: "Oh, what fools we have been; we ought to have seen that before."[5] By chance, Bohr was preparing to leave for America in early January to attend an international conference on theoretical physics at George Washington University in Washington, DC. There, Bohr's announcement of the new discovery so galvanized the scientific gathering that many of the physicists raced back to their laboratories to repeat the experiment for themselves.

By early 1939, the ancient dream of the alchemist – changing one element to another – had become reality. Even more startling, the scientists discovered that they had liberated the enormous power of the atom. If, as suspected, each fissioned nucleus released more than one neutron (the average later turned out to be 2.5), then a self-sustaining chain reaction might be possible. Within a year, over one hundred scientific papers on this matter had broken into print.[6]

Once the facts were established, the scientists sought to explain the theory behind them. Niels Bohr concluded that it was not the heavy uranium isotope U^{238} that had fissioned. Instead, the fission process had occurred in the rare isotope, U^{235}, which composed only about 0.7 per cent of uranium in its natural state, and was scattered equally throughout the material. Bohr's interpretation explained why the natural concentrations of uranium had not exploded spontaneously.[7]

Although the physicists immediately realized the *theoretical* possibility of building a nuclear weapon, in 1939/40 the majority felt that it would never be *practical*. It was believed that it would take perhaps 180 tons of uranium to explode in any fission bomb. The major names in the field of physics all shared this opinion. A few years earlier, in a famous gaffe, Nobel Laureate Lord Rutherford had dismissed the release of atomic energy as "moonshine." In 1939, John Anderson, then Lord Privy Seal, termed an atomic bomb "a scientific but remote possibility." Einstein and Bohr felt the same way. Bohr argued that it would take the entire efforts of a nation to build a bomb, a position he held until 1943. From London, Winston Churchill suggested that widespread rumors of a "super" weapon might simply be a German attempt at blackmail.[8]

The German invasion of Poland on September 2, 1939, made the question more than academic. Seventeen days later, on September 19, 1939, Hitler drove to Danzig to make a dramatic speech. There he threatened France and England with "a weapon against which there is no defense." British intelligence officers monitored Hitler's comments with care and puzzled over what the German Chancellor had meant. They came up with several possibilities: Hitler might have been bluffing. He might have been referring to a new poison gas. Or, most likely in retrospect, he might have been pointing to the power of the Luftwaffe. Still, in the fall of 1939, no possibility could safely be excluded. The German expertise in nuclear research was well known.[9]

Thus, in early 1940 Sir Henry Tizard, Chairman of the Committee on the Scientific Survey of Air Defence, gave the go-ahead for British scientists to investigate the possibility of a nuclear weapon. He also established a committee of well-known British Jews to interview the Continental Jewish refugee scientists to see what they could discover. The official historian of the British Intelligence Services, Francis Harry Hinsley, has noted that in 1939/40, the British could neither confirm nor deny the existence of a German atomic bomb. Science writer Ronald Clark has suggested that if the British intelligence penetration of Germany had been more sophisticated, no bomb would have been developed by August of 1945.[10]

Since many of the Jewish refugee scientists had found homes in Britain, it seems logical to assume that they might have had special impetus to work on the British atomic bomb project. As Rudolf Peierls later recalled, however, it was largely force of circumstance, not revenge against the Nazis, that drew the refugee scientists into the fledgling British atomic program. Faced with the immediacy of German air raids, the native-born British scientists were all enlisted to work on radar and coastal defenses. Large numbers left their laboratories during the first week of the war, effectively abandoning nuclear physics for about seven or eight months. They renewed their interest only when they heard rumors that all German scientists had been summoned to the Kaiser Wilhelm Institute in Berlin.[11]

The refugee scientists, most of them of Hungarian, German or Austrian origin (and not yet British citizens), were excluded from radar and harbor defense work. In the eyes of the government, they were "enemy aliens." For example, when Polish émigré Joseph Rotblat initially offered his services as a physicist for the British war effort, he was politely refused.[12] Thus, a disproportionately large number of Jewish émigré scientists found themselves engaged in preliminary nuclear research on what was considered to be the remote possibility of an atomic weapon. Tizard even grumbled that to allow the refugee scientists to utilize uranium when the English physicists were engaged in other war work might give them unfair advantage later.

Soon Rudolf Peierls and Otto Frisch, two of the foremost refugees, began to work on this question in earnest. Born in a suburb of Berlin, Peierls was in England on a Rockefeller Foundation Fellowship when the Nazis came to power and decided not to return. A member of the celebrated "class of 1933," Peierls assumed the position of an honorary research fellow at the University of Manchester in 1934. Five years later, he had risen to Professor of Applied Mathematics at the University of Birmingham, even though he was only 30. He became a British citizen in 1940. Frisch, three years his senior, had taken a teaching position at Birmingham in the summer of 1939, and still remained a citizen of Germany.

Peierls had special interest in problems of critical mass while Frisch was particularly concerned with isotope separation. Excluded from other war work, the two men began sharing ideas of nuclear fission with each other. In March of 1940, the two scientists combined to produce one of the lesser known but most vital scientific documents of the early war years: The Frisch-Peierls Memorandum. (See Appendix II.)

The three-page Memorandum contained two parts. The first was a technical blueprint for a potential atomic weapon. Here the two men countered conventional wisdom by suggesting that the amount of fissionable material (U^{235}) needed for an atomic bomb would be far less than previously expected. (Their initial estimate was under a pound; later they revised it upward to about five kilograms.) Britain could utilize commonly known industrial isotope separation techniques to produce the material. The only problem might be cost.

Part two of the Memorandum set forth a discussion of the strategy of use. Frisch and Peierls were the first people to raise the question of radioactive fall-out and the first to query the morality of using an atomic weapon. Indeed, Frisch and Peierls suggested that because of the indiscriminate destruction that such a weapon would necessarily entail, the British might find it morally repugnant. The average person has never heard of the Frisch-Peierls Memorandum, but it ranks as one of the most significant scientific documents of the twentieth century. Perhaps the best way to comprehend its importance is to contemplate what might have happened "if Frisch and Peierls had composed it in their native Germany and delivered it to a German government."[13] As Margaret Gowing has noted, Frisch and Peierls performed one of the most difficult tasks in the historical development of science: "They had asked the right questions."[14]

Shortly afterwards, Churchill's coalition government organized a special committee of scientists to study this problem in detail. Since the group was headed by G.P. Thomson, Professor of Physics at Imperial College, London, it was known initially as the "Thomson Committee." However, it is much more famous as the "Maud Committee," a renaming stemming from one of the most famous anecdotes of the early war years. Shortly after the Germans overran Denmark, British physicist Dr John Cockcroft received a message concerning the welfare of Niels and Margrethe Bohr from Lise Meitner in Sweden: "Met Niels and Margrethe recently. Both well but unhappy about events. Please inform Cockcroft and Maud Ray Kent." Puzzled by the "Maud Ray Kent" reference, Cockcroft showed the message to British Military Intelligence. One cryptoanalyst substituted "i" for "y" to produce an anagram: "radium taken." Another suggested that it was an anagram for "Make Ur Day Nt [Make Uranium Day and Night]." This was "all very wild," a friend wrote Sir Henry Tizard, "but just sufficiently reasonable to make one worry." Only when Bohr arrived in England in 1943 did the British discover that Bohr had simply asked about the woman who had been governess to his children, Maud Ray, who lived in Kent.[15]

Margaret Gowing has described the Maud Committee as "one of the most effective Committees that ever existed." Immediately it began to conduct a series of experiments to elaborate on the suggestions of the Frisch-Peierls Memorandum.[16] The Committee conferred for fifteen months, issuing its final reports only in June and July of 1941.

One of the final Maud reports dealt with the potential use of nuclear power. Another considered the timetable of weapons development. The Maud Committee estimated that it would be possible to make a bomb out of U^{235} by the end of 1943 and concluded that it would definitely work. They recommended that specific technical measures be initiated immediately. All the members of the Maud Committee felt that the effort would be worthwhile, Peierls recalled later, even though the project "might cost as much as a battleship."[17] The Committee also noted that even if the war ended without the use of a bomb, "The effort would not be wasted, except in the unlikely event of complete disarmament, since no nation would care to risk being caught without a weapon of such decisive possibilities."[18] From the outset, then, the British government had one eye cast on the probable nuclear politics of the postwar world.

The Frisch-Peierls Memorandum and the Maud Report were milestones in the race for the secret of an atomic weapon. Hungarian émigré physicist Leo Szilard always credited the British with being the first to recognize that one could separate enough U^{235} for a weapon and for alerting the Americans to this possibility. J. Robert Oppenheimer later admitted that the Maud Report transformed the American program from a series of desultory committees to a focused, concentrated effort. Historian A.J.R. Groom has estimated that the Maud Report accelerated the United States' weapons program by a minimum of six months.[19] Margaret Gowing stated it even more forcefully: "Without it, [the Maud Report]," she wrote, "World War II would almost certainly have ended before an atomic bomb was dropped.

Britain's conception of herself as a nuclear power was born out of the Frisch-Peierls Memorandum and the Maud Committee Report."[20]

On August 30, 1941, Prime Minister Winston Churchill gave his assent to further research in his famous minute to the Chief of Staff: "Although personally I am quite content with the existing explosives, I feel we must not stand in the path of improvement . . ."[21] Afterwards the Maud Committee disbanded, and the British began their atomic weapons program in earnest in October of 1941. They placed it under the Department of Scientific and Industrial Research (DSIR), which was headed by Sir John Anderson, Lord President of the Council.

The program was code named "Tube Alloys." Wallace Akers, the director of Atomic Energy Research, and his deputy, Michael Perrin (both from ICI), had first suggested the code name of "Tank Alloys." The term had no meaning, but since tanks were so vital to the war effort, it indicated a top priority. Anderson, however, demurred. He argued that tanks might not retain their top priority throughout the entire conflict, and suggested that tubes, an item found in virtually everything connected with the war, would be a more suitable term. Thus "Tube Alloys," or TA (English, "tewb," American, "toob") became the official designation for the British atomic bomb project.[22]

From 1941 to 1945, the atomic program was known by various cover names. "Tube Alloys," of course, had little meaning for Americans. In October, 1941, President Franklin Roosevelt wrote Churchill suggesting the code name "Maudson," later changed to "Mayson" (out of fear that "Maud" had been compromised) but that never caught on. The program was occasionally termed the "DSM Project" and Secretary of War Henry Stimson always referred to it in his diary as "S-1" (Section 1 of the Office of Scientific Research and Development). In August, 1942, the Army Corps of Engineers established a special engineer district, one without geographical limits, that was headquartered in Manhattan. Thus, the terms "Manhattan Engineer District" (MED) or "Manhattan Project" came into being. This eventually became the most common name.[23]

From 1940 to 1943, Anglo-American scientific cooperation on nuclear matters went through three distinct stages. With the outbreak of the war, Winston Churchill realized that the United States and Great Britain had to forge an "Anglo-Saxon bloc," if the British were to escape the Nazis. During the early years of conflict, Britain was willing to do almost anything to ensure American cooperation. The fall of France in June of 1940 only heightened Churchill's concern. By mid-1941, his coalition government had sent perhaps 3,000 British officers, businessmen and Foreign Office officials to Washington, DC to promote the British cause. A year later, the "Washington Whitehall" had tripled.[24]

Numerous American governmental officials shared Churchill's fears. This can be seen on a personal level by the extensive Churchill-Roosevelt correspondence. On the national level, it is reflected in the Lend Lease Acts of 1940/41, so ably engineered by British Supply Council agent Arthur Purvis. On the scientific level, the collaboration may be best illustrated by the Fall, 1940, visit to the States by Sir Henry Tizard. As early as November, 1939, Tizard had expressed a desire "to bring American scientists into the war before their government," but it took the British military disasters of 1940 before Churchill allowed him to work on the idea. Rejecting the Admiralty's position that all technical information be exchanged with the Americans only on a *quid pro quo* basis, Churchill told Tizard to withhold nothing from American scientists.

Thus, on August 31, 1940, Tizard set sail for Washington with the famous "black box" of the latest British scientific and technical secrets. Items in the box included: the design of the Rolls-Royce Merlin engine, which would later be extensively used in the P-50 Mustang; plans for power-driven turrets and anti-submarine devices; the proximity fuse; a gun-laying predictor for anti-aircraft work; new information on chemical weapons; advanced data on jet engines; and the cavity magnetron, which would be vital for the development of American radar.[25] (As it turned out, the invention of microwave radar became the most significant weapon the Germans never developed.)[26] This exchange of radar information, one scientist later remarked, would be the "most famous example of a reverse lend lease."[27]

Equally important, physicist Sir John Cockcroft accompanied Tizard on this journey. He shared the latest British accomplishments in the field of nuclear research. Thus, by late 1940, scientists from both nations had established a fairly free exchange of nuclear information.[28] When visiting American physicists toured the British laboratories, they were impressed by how seriously the British were taking the nuclear question. "I wish I could tell you that the bomb is not going to work," British physicist James Chadwick told two American scientists, "but I am 90 per cent certain that it will."[29] In April, 1941, Harvard physicist Kenneth T. Bainbridge visited England as a representative of the National Defense Research Committee (NDRC). Upon his return he relayed the British sense of urgency to Vannevar Bush, director of the Office of Scientific Research and Development (OSRD) and James Conant, Chairman of the NDRC. The British were convinced that the Germans had a head start in this area, and that time was of the utmost importance.[30]

In August, 1941, Australian Marcus Oliphant sailed to the States to discuss various radar matters. But he also carried a letter from George Thomson urging him to inquire why the American "Uranium Committee," headed by Lyman Briggs of the National Bureau of Standards, had made no reply after receiving the Maud Committee report. When Oliphant met Briggs, he discovered that Briggs had not read the Maud report thoroughly. Moreover, he had sequestered the top secret document in his safe so that no other American scientists had read it either!

Incensed at this delay, the volatile Oliphant began badgering all the American scientists he could find, including James Conant, Vannevar Bush, Enrico Fermi, and, especially, Ernest O. Lawrence of Berkeley. Time was crucial, Oliphant repeatedly said, for the Germans were obviously ahead of both the British and the Americans.[31] Leo Szilard was so impressed with Oliphant's private crusade that he later suggested that Congress create a special medal for Marcus Oliphant: one that commemorated "distinguished services" by "meddling foreigners."[32]

In 1942, a four-man British delegation also visited the States to discuss further cooperation. When Rudolf Peierls met Physicists Arthur Compton and Robert Oppenheimer, he complained that America had not given sufficient thought and energy to the matter at hand. Peierls also criticized both the Uranium Committee and the lack of effort by Yale Physicist Gregory Breit, then coordinator of fission with fast neutrons ("the coordinator of rapid rupture").[33] In turn, however, the Americans complained that the British scientists behaved in a condescending fashion. Overall, they made a rather poor impression on the Americans.[34]

Bush and Conant utilized the mounting consensus to spur further American action. They took several steps toward streamlining the committee process.[35] Until the Japanese attacked Pearl Harbor, however, the American atomic effort lacked the intensity of the British "Tube Alloys" Program.

All through 1941, the Americans saw considerable advantage to continued cooperation with the British on uranium matters. In October, Roosevelt wrote Churchill to that effect. But Churchill's government suddenly began to cool to the American attempt to strengthen the US–UK nuclear alliance. Presumably, the British felt the Americans at that time could add little to their program. For reasons that are not entirely clear, the British began to resist American attempts at stronger integration of the two scientific programs. This rebuff was to have serious, long-term consequences. As historian Barton J. Bernstein has observed, the lost opportunity would prove to be a costly error on the part of the British.[36] Margaret Gowing put it even more bluntly: "They [the British] missed the bus."[37]

The next major phase of the early US–UK nuclear exchange occurred during mid-1942, when the Americans had seriously begun to mobilize on their own. With America's entrance into the war, however, Anglo-American nuclear cooperation became much more strained. The British blamed the Army, which had assumed control of the Manhattan Project and, especially, the irascible General Leslie R. Groves, who took over as project head on September 17, 1942. Groves restricted US–UK contact considerably. Later he confessed he had always distrusted the British; but, in fairness, Groves distrusted virtually everybody.

The final phase of the US–UK nuclear wartime relationship began during the summer of 1942. Faced with constant air raids and manpower and material shortages, the British changed tactics once again. Their earlier enthusiasm for building a bomb had dwindled with the realization that they would have to merge their program with the Americans. Visits to American installations confirmed the growing gap between the two projects. "We must face the fact," Sir John Anderson wrote Churchill on July 30, 1942, "that . . . [our] pioneering work . . . is a dwindling asset and that, unless we capitalize it quickly, we shall be outstepped. We now have a real contribution to make to a 'merger.' Soon we shall have little or none."[38]

By the fall of 1942, America's gigantic industrial facilities had come into play. So, too, had Groves' zeal on questions of security. The General's basic principle of operation was the Army's "need to know" rule. This meant that a person could have access to scientific or technical data only if this information could be proven essential to the job at hand. Such "compartmentalization" of information contrasted sharply with the open atmosphere of scientific research in Great Britain. The assumption there was, the more information people had, the better.[39] Compartmentalization also completely excluded the British scientists from the American program. W.A. Akers strongly protested at the exclusion of British scientists from Manhattan Project data, but to no avail. His colleague, James Chadwick, argued that the American reluctance to include British personnel had two main causes. First, many of the British scientists were foreign born, most of them refugees from Nazi-occupied countries, and this made them slightly suspect in the eyes of American security. Second, Groves and the Americans had become suspicious that the giants of British industry, especially ICI, had already made plans to corner the postwar nuclear energy market.[40] In addition, Akers felt the overall atmosphere was hampered by the "general feeling prevalent in America that the British always got the best of any deal."[41]

There was some truth in these fears. The Americans kept close watch over their own "enemy alien" scientists, and several of Roosevelt's advisors, including both Conant and Bush, maintained that if Roosevelt made any agreement with the British that dealt chiefly with postwar nuclear power, he would be exceeding the limits of the War Powers Act. The Americans also argued that if compartmentalization kept valuable information away from large American companies, how could they legitimately share this data with scientists from another nation?

Successful transatlantic experiments with plutonium, element 94, added yet another factor to the equation. As early as 1940, Cambridge physicist Egon Bretscher had speculated as to the explosive potential of plutonium. The Maud Report also discussed this possibility. Then, in 1941/42 Berkeley physicists Glen Seaborg and Emilio Segré successfully isolated plutonium. Because of the chemical differences, separating plutonium from uranium 238 was considerably easier than separating U^{235} from U^{238}. The American scientists showed little interest in sharing their plutonium breakthroughs with the British.

Finally, American spokesmen distrusted British industry in general and former ICI engineer W.A. Akers in particular.[42] The end result was that British–American relations in nuclear matters remained very strained for almost two years. Historian Andrew Pierre has observed that there was almost as much conflict as cooperation on this issue during the early wartime period.[43] Historians Richard G. Hewlett and Oscar E. Anderson, Jr. have argued that the quarrel over nuclear matters formed one of the most intricate and divisive issues in the long annals of Anglo-American relations.[44] The dispute thus ranked with the 1783 Treaty of Paris, the termination of the War of 1812, or the *Alabama* claims after the Civil War.

A chief area of distrust lay with each side's suspicion of the other's plans for the postwar world. With centuries of experience in postwar settlements, the British held a more sophisticated view of probable power relationships after the conflict had ended. The Maud Committee had concluded that no nation could be secure in a postwar world without atomic weapons. Sir John Anderson once told Canadian Prime Minister Mackenzie King that "an absolute control" would be given to whatever country possessed the "secret."[45]

The British felt so strongly about this that in May of 1943 they threatened to risk a delay in the quest for victory in Europe by diverting scientific manpower to the production of the bomb. They stated frankly to several American officials that their major concern was possession of the bomb in the postwar world.[46] While nuclear power stations were never completely discounted, this was decidedly a secondary issue to that of nuclear weapons.[47]

The Americans did not see it this way. They voiced deep suspicion of the potential *commercial* use that the British would make of the atomic "secret." As late as 1965, Vannevar Bush claimed that the chief reason Churchill pushed Roosevelt on this matter was his desire to create atomic energy plants to solve Britain's expected postwar coal problem.[48] A few historians have also echoed that position.[49]

Prime Minister Churchill reacted strongly to the 1942/43 impasse in nuclear relations. He repeatedly tried to convince Roosevelt of the legitimacy of British claims. Initially, his personal diplomacy seemed to work. When the two Allied leaders met at Hyde Park, New York, on June 20, 1942 to discuss the sharing of atomic weapons information, the atmosphere seemed most cordial. Churchill later recalled that "the whole basis of the conversation was that there was to be complete cooperation and sharing of results."[50] Still, the expected cooperation was slow to develop from the American side.

The meeting at Casablanca in January of 1943 allowed Churchill an opportunity to prod again the President's Special Assistant, Harry Hopkins ("Lord Root of the Matter"), on the issue. Churchill was outraged that the American War Department wanted knowledge of the British nuclear expertise but refused to divulge any of theirs.[51] After the conference, Churchill sent a steady string of cables to Hopkins on this matter. The proposal to build a nuclear production plant in North America, he reminded Hopkins, was "not due to any technical inability on the part of the British." Rather, it was a purely strategic move.[52] When Roosevelt turned to Hopkins for advice in July of 1943, Hopkins supported Churchill: the President had made a firm commitment to Churchill on the matter of exchange, Hopkins stated. The US should follow through with it.[53]

The British scientists were equally concerned about the exclusion. One wrote to M.W. Perrin in London that the American project could probably get along without British presence except, perhaps, in the gaseous diffusion method of isotope separation. "We can indeed help them accelerate their programme, perhaps considerably," he noted, "but nothing more."[54] From his office in Washington, Akers argued that it was essential that the Americans believe that the British could assist them *during the war itself.* If the Americans concluded that the British could not be of immediate help, he wrote, they would obviously be reluctant to share nuclear knowledge that would have only postwar application.[55]

This disagreement was finally resolved in the Quebec Conference of August, 1943. Historian B.L. Villa has suggested that the Quebec negotiations revealed Roosevelt at his diplomatic best. In Villa's view, the American President traded atomic cooperation with the British in return for Winston Churchill's agreement for a D-Day cross-Channel invasion of Europe during the next year. This proved no easy task. Roosevelt had to overcome both Churchill's fears that the English Channel would be floating with corpses and his own scientific advisors' claims that there was no need to tell the British anything about the rapidly accelerating American nuclear program.[56]

This Quebec Agreement established the official nuclear relationship between the nations for the duration of the war. The two countries agreed never to use the weapon against each other; not to use it against a third party without mutual consent; not to communicate information to a third party without mutual consent and that:

> In view of the heavy burden of production, the British government recognize that any post-war advantages of an industrial or commercial character shall be dealt with as between the United States and Great Britain on terms to be specified by the President of the United States to the Prime Minister of Great Britain. The Prime Minister expressly disclaims any interest in these industrial and commercial aspects beyond what may be considered by the President of the United States to be fair and just and in harmony with the economic welfare of the world.[57]

The Quebec Agreement further stated that the two countries should combine their uranium efforts for a massive joint project. Roosevelt based his decision on the assumption that the weapon could be developed *before hostilities ended*. Because he so assumed, Anglo-American nuclear cooperation fell within the general agreements covering the interchange of technical secrets on research and inventions. British scientists would cross the Atlantic to work in the States because America offered both a respite from German bombing raids as well as untapped industrial might. A Combined Policy Committee based in Washington would oversee all US–UK (and Canadian) efforts.[58] Wallace A. Akers, whom Bush described as "a very able man, but not the one to handle the matter,"[59] was declared *persona non grata*. He would be replaced by James Chadwick. In effect, the British relinquished their independent bomb effort to become a "junior partner" in the American enterprise.[60] Groves later credited the smooth workings of the Combined Policy Committee to the fact that the British realized their contribution to the Manhattan Project was minimal.[61]

In retrospect, the Quebec Agreement proved the best possible compromise under the circumstances. Britain clearly lacked the resources to carry the project to completion. As Churchill recalled in his memoirs, "it seemed impossible to erect on the Island the vast and conspicuous factories that were needed." Yet the nation that had first produced the Frisch-Peierls Memorandum could hardly relinquish the operation either. Thus, Churchill's government assumed the "bridesmaid's" role as the only available course.[62]

With these constraints in mind, British Mission scientists relocated to the United States. In fact, the day after the Quebec Agreement was signed, James Chadwick, Francis Simon, Marcus Oliphant, and Rudolf Peierls arrived in Washington, "with almost indecent haste," the Americans thought.[63]

Groves immediately laid plans to utilize British expertise in the gaseous diffusion method of separating U^{235} from U^{238}. British scientists had initiated this process, but Columbia University physicist Harold Urey expressed grave doubts that their methods would work. Consequently, Groves arranged for a fifteen-member British delegation to meet with American experts in New York City in late December, 1943. On January 5, 1944, the two sides exchanged views in an intense, four-hour meeting.

The issues discussed were crucial, for they would determine the design, process, and barrier system for the uranium separation facility then under construction in Oak Ridge, Tennessee. The Americans opted for a newly invented barrier process, but the British argued that such changes would delay U^{235} production, perhaps until the summer of 1946. After considerable thought, Groves decided to back the American plan.[64]

During this conference, Groves also drew on Rudolf Peierls' suggestions as to how British talent could best be utilized in the various American installations. Afterwards, Groves assigned about thirty scientists, including Oliphant, to work with American physicist E.O. Lawrence in Berkeley on electromagnetic isotope separation. He sent about thirty others to laboratories in Oak Ridge, and to various locations in Canada. (The program at Montreal and, later, Chalk River, was a British-Canadian-French project that concentrated chiefly on postwar power production.) A few, such as Peierls, Fuchs, and Skyrme, remained in New York City.

Meanwhile, Groves' insistence on the "need to know" policy had frustrated more than just the British. Compartmentalization also hampered the necessary interaction between American scientists in various areas of the project. Since all the problems were interconnected, they argued that compartmentalization actually hindered the development of the weapon. (After the war Leo Szilard stoutly maintained that Groves' insistence on compartmentalization had held up the project by as much as eighteen months.)[65]

Groves never gave way completely, but he did bow with the prevailing winds. Under the pressure, he agreed to create a new installation – termed site Y – where communication could be more open. This new site had to be located in the interior of the country far away from potential attack. It also had to be secure from espionage.

J. Robert Oppenheimer, the General's choice to head the new installation, had known the area of northern New Mexico from his youth. His family owned a ranch in the rugged Pecos mountains, just north of Santa Fe. Oppenheimer convinced Groves that New Mexico would be the ideal location for the new laboratory.

In the fall of 1942, Groves and Oppenheimer toured the region to select the location. Some sentiment was expressed for the hamlet of Jemez Springs, about 50 miles from Santa Fe, but Oppenheimer balked. He felt that "the surrounding cliffs would give his people claustrophobia" and that the valley was too narrow for any unexpected expansion of the project.[66] Thus they turned to a second choice, an exclusive boys' school – the Los Alamos Ranch School – located at the top of a prominent mesa. When the search team saw the site, they concurred. The Ranch School was closed and the Army arrived *en masse* in late 1942 to take over the property.

The selection of Los Alamos proved a wise one. About thirty-five miles northwest of Santa Fe, the site lay on a 7,000-foot mesa composed of volcanic tuff. A few miles to the west lay the Valle Grande, a gigantic caldera, which, ironically, had probably produced the largest explosion in North America when it erupted in prehistoric times. Access to the community could be carefully controlled through the one two-lane road that snaked up the valley to the mesa. Armed security guards on horseback also guarded the region. K-9 (canine) Corps dogs patrolled the base of the cliffs.[67] The site lay near ancient Anasazi ruins, several thriving Indian pueblos, and numerous isolated Hispanic villages. Albuquerque, locus of the main airport for the region, lay only 90 miles away.

After the site had been selected, Oppenheimer began recruiting his team of scientists in earnest. Several, however, balked at the prospect of coming to an Army camp where military discipline, rather than a scientific give-and-take, would reign supreme. The initial plan was to draft all the scientists in the Army, but I.I. Rabi persuaded Oppenheimer to maintain Los Alamos as a civilian outpost. This decision made recruiting much easier, and Oppenheimer's secret project soon decimated the physics departments of Harvard, Princeton, Illinois, Cornell, and elsewhere. In 1944, the completion of several types of radar projects at MIT also freed numerous MIT scientists to go west to join them. From a few hundred scientists and their families in 1943, Los Alamos grew to over 7,000 people in 1945.

Located in one of the most scenic spots of the nation, Los Alamos slowly began to develop into a community like no other. It became "celebrity land," "a concentration camp for Nobel Laureates," or "the most super-charged intellectual atmosphere that had ever existed." What Hollywood was to the aspiring starlet, one resident remarked, Los Alamos was to the world of science.[68] Virtually all the major names in the physics community were there – Hans Bethe, Victor Weisskopf, Stanislaw Ulam, Enrico Fermi, Edward Teller, John von Neumann, Emilio Segré, Robert Wilson, plus a number of younger men, such as Richard Feynman, who would later achieve greatness in their fields. In the twenty-seven months left of wartime, Los Alamos provided a unique moment in the history of the human race. It was the nation's "Athenian Age." This moment, moreover, was considerably enhanced by the arrival of the twenty scientists from the British Mission.

2 The British Mission at Los Alamos: The Scientific Dimension

The first members of the British Mission – Otto Frisch and Ernest and Peggy Titterton – arrived in Los Alamos in mid-December, 1943, soon to be followed by Sir James and Lady Chadwick. Egon Bretscher came in February, 1944 and Rudolf and Genia Peierls shortly thereafter. The team was essentially complete by the fall of 1944, although Canadians Kathleen and J. Carson Mark did not arrive until May of 1945.[1]

In 1943–4, the British émigrés came from a land still staggering under the burden of wartime rationing. Regulations enacted during the fall of 1942 limited each citizen to one shilling and twopence worth of meat per week. Adults received two pints of milk a week in winter and four to five pints in summer. Eggs were limited to thirty a year and cheese to just a few ounces per person, per week, although children were allowed more. White bread became an almost impossible luxury, and the government fined people for feeding bread to the birds. Although few people actually starved – potatoes, bread, and cabbage were never rationed during the war – the diet had a dreary sameness to it. Gasoline became almost unavailable, and heating oils were carefully guarded. Clothes rationing cut the heart out of fashion, and the war introduced a drab, plain style of British dress that continued for decades afterwards. "Make do and mend" became a popular saying among housewives. At the height of rationing, it was not uncommon for chemist shops to put a list in the window of items they did *not* have.[2]

Coming from this environnment, many of the British scientists were astounded by the cornucopia of the United States. A.P. French remembered his first impression of America as "a land of milk and honey." Michael Poole recalled his move as one from "total shortages to total plenty." French and Poole were both late in reporting to the British Embassy in Washington because they wandered for two hours looking in the shop windows. Donald G. Marshall described his first views of the lights of New York City as "absolute magic."

The shift from British food rationing to American abundance often proved memorable. When Derrick Littler and Harold Sheard missed a flight connection in Washington, they eagerly utilized their vouchers to order

16

large meals. When the various dishes arrived, they first thought they were only to help themselves from each one. Moreover, after the *hors d'oeuvres*, they found, to their dismay, that they could eat no more – so much had their stomachs shrunk from British austerity. Otto Frisch actually let out a squeal of delight when he spied his first fresh oranges for sale at a Norfolk fruitstand.[3]

Tony Skyrme had a somewhat ruder introduction to American life. Newly arrived in New York, Skyrme had embarked on a Friday evening stroll through Central Park when he was stopped by a police patrol searching for draft evaders. Because he could not produce the proper papers to prove his draft exemption, and had what was described as a "gutteral accent," Skyrme was suspected of being a spy. The patrol placed him in temporary custody and subsequently transferred him to a federal prison. There he fell in contact with another incarcerated Englishman. "Do you remember the Air Raid Shelter Murders in 1940?" the man asked Skyrme. "That was me." When Monday arrived, Skyrme was released and sent back to work.[4]

The trip to Los Alamos itself – the tortuous drive up the winding road to the mesa top, the magnificent scenery, the ever-present mud and construction – always proved startling to the British. But arrival on the Hill was usually enhanced by greetings from old friends, all of whom wanted the latest news from the UK. When A.P. French, Michael J. Poole and James Hughes arrived in Los Alamos in October of 1944, they were met by their senior colleague, Philip B. Moon, who served as informal "secretary." Moon directed them to Robert Serber's "Los Alamos Primer," a basic typescript introduction to the Manhattan Project, which they had to read, cover to cover, within the restricted Technical Area. Over forty years later, French still vividly recalled his first reading of Serber's "Primer."

The young men – all fresh from Cambridge, Birmingham or Liverpool – were slightly overwhelmed at first. None of them had heard of the Frisch-Peierls Memorandum or Tube Alloys or engaged in previous discussion of the actual design of the bomb. Los Alamos made their earlier efforts seem small scale, French recalled, because "at Cambridge we had only these very academic measurements of cross sections and so on." The Los Alamos scientists had moved the discussion out of the realm of theory. It was clear that everyone there expected that the bomb would soon become a reality.[5]

For several members of the Mission, the transfer to Los Alamos proved less startling. They arrived on the Hill after several months of work at other North American Manhattan Project sites. George Placzek, for example, came to New Mexico from the Chalk River establishment in Montreal. Peierls, Fuchs and Skyrme all arrived from New York City, where they had worked on the problems of gaseous diffusion. Consequently, they were a little more "acclimatized" to American ways than those who arrived fresh from the British Isles.

Several months earlier, Michael Perrin had noted that it was crucial that the British transfer to the USA "as good and complete a team as possible." This was necessary for two reasons – both to speed up the timescale of the Manhattan Project and to bring back valuable experience for future nuclear affairs.[6]

The high calibre of the scientists reflected Perrin's selection policy. The British Mission included a most impressive array of scientific and technical expertise. With the possible exception of Albert Einstein, Niels Bohr was widely considered the premier mind in the world of physics. James Chadwick was a Nobel Laureate. Otto Frisch and Rudolf Peierls were both rising rapidly in the profession, while Swiss émigré Egon Bretscher, perhaps the first man to foresee that plutonium could be used as an energy source, was only a half-step behind. W. Gregory Marley ranked as one of the world's foremost photographic experts. J. Carson Mark was described as one of the "ablest men" ever to graduate from the University of Western Ontario.[7] G.I. Taylor and William G. Penney would both later be showered with scientific and royal honors. P.B. Moon from Birmingham University had already served on the prestigious Maud Committee. Several of the more junior people, such as Tony Skyrme, D.J. Littler, D.G. Marshall, A.P. French, and Michael Poole, would all go on to distinguished scientific careers. The overall calibre of the British Mission to Los Alamos may be estimated by the fact that six men – Frisch, Peierls, Bretscher, Moon, and Placzek – became heads of Los Alamos groups. Gregory Marley headed a section. The Mission included seven experimental physicists, five theoretical physicists, two electronics experts, and five specialists on explosives. It was quite a team.

The official head of the British Mission at Los Alamos was James Chadwick. A colleague of Lord Rutherford's at Cambridge, Chadwick had discovered the neutron in 1932, ("one of the great scientific discoveries of all time"), for which he received the Nobel Prize for Physics in 1935. He also replaced Akers on the Combined Policy Committee. The initial plan was to have the British scientists all work directly under Chadwick, on problems of his selection.[8] Groves favored this scheme and Oppenheimer, perhaps under pressure, went along.[9] Within a short time, however, this idea was rejected in favor of full integration of the British team into all divisions of the Laboratory.

Thus, members of the British Mission were assigned to a variety of tasks. Over the two-year period, the British scientists worked in nearly all of the existing Los Alamos Laboratory divisions. The only dimension of the

project they were systematically excluded from was the chemistry of plutonium. Only Chadwick ever visited the plutonium production plants at
Hanford, Washington. But when discussions of plutonium production arose
in the Los Alamos weekly Tuesday colloquium meetings, as they regularly
did during the Spring of 1945, the British scientists were never sent from
the room. An initial attempt by American security to monitor all the scientific
reports that the British scientists utilized soon fell into disarray.[10] Thus, the
British Mission members were fully integrated into all Los Alamos Laboratory activities. Their white (top clearance) Laboratory badges meant
they could enter any area behind the fence.

Their assignment was straightforward: the gigantic plants at Hanford
and Oak Ridge would produce the necessary plutonium and enriched uranium. The scientists at Los Alamos had to devise a weapons system for
each material. It was not long before they discovered that each weapon
demanded a very different format. The U^{235} device proved relatively simple.
It consisted of firing two sub-critical masses of enriched uranium toward
each other in a "gun assembly." When the two masses met, they went
critical and exploded. This was the weapon first conceived by Frisch and
Peierls in their Memorandum. In Liverpool, Chadwick had independently
reached the same conclusion. Because of the shortage of U^{235} fissionable
materials and because the scientists were positive of the theory, the "gun
assembly" weapon was never field tested. It became the Hiroshima bomb.

The plutonium device proved far more complicated. Although plutonium
was easier to manufacture than U^{235}, it could not be detonated in the same
"gun assembly" fashion. If two sub-critical pieces of plutonium had been
fired at one another, they would have detonated prematurely. Thus, the Los
Alamos scientists had to devise another means to detonate a plutonium
bomb. (They could not wait for Oak Ridge and Berkeley to produce enough
U^{235} for a second weapon.)

After much experimentation, the solution to this problem emerged as the
theory of "implosion." Implosion involved the placement of a twelve pound
sphere of plutonium in the center of conventional explosives. These explosives were detonated with microsecond precision to "squeeze" the plutonium down to the size of a walnut. With that, the sphere would go critical.
Because of the complexity, the Army decided to field test the weapon
before combat use. This test took place at Trinity Site, New Mexico on July
16, 1945. A similar implosion weapon was dropped on Nagasaki on August
9, 1945.

One of J. Robert Oppenheimer's most critical early appointments came
when he selected Cornell physicist Hans Bethe as head of the Theoretical
(T) Division, the most prestigious section of the Laboratory. In so doing,

however, he had to pass over the equally qualified Edward Teller and Felix Bloch, both of whom harbored considerable resentment. Bloch eventually left Los Alamos, and Teller became increasingly difficult to work with. When Bethe decided that they should concentrate on the implosion system as "the most important task in the theoretical division," Teller refused to cooperate. He insisted on working on his own project, the "super" or hydrogen bomb. Thus, Oppenheimer had a delicate personality issue on his hands. He resolved it in June, 1944 by assigning Teller his own group (within Fermi's F Division), which reported directly to Oppenheimer.

The loss of Teller, however, seriously weakened Bethe's theoretical team. Oppenheimer had worried about the strength of the American theoretical group since the fall of 1943. After Teller's departure, the director wrote Groves that Rudolf Peierls would have to replace Teller in that configuration.[11] Shortly afterwards, Oppenheimer lauded Peierls' contributions to the hydrodynamics of implosion.[12]

In defense of Teller, the quest for a hydrogen weapon also fell within the initial Manhattan Project parameters of a nuclear weapon "within the shortest possible time." From 1942 forward, the possibility of producing a hydrogen bomb before the end of the conflict had both risen and fallen. In 1942, several men believed that going from a fission weapon to a hydrogen weapon would not be too difficult a task. The construction of a hydrogen liquefaction plant at Los Alamos, one of the first structures erected, reflected that optimism.[13] By 1944, however, such hopes had dimmed. The brunt of the Los Alamos effort was always directed toward the fission weapons, which were clearly possible.[14]

The Teller–Oppenheimer tension that began at Los Alamos would later divide the American scientific community almost in twain. The problems were both personal and professional, and every minor incident contained the potential for confrontation. This may be seen in the following story: in late 1944, Lord Cherwell, Churchill's science advisor, visited the laboratory for a day. Oppenheimer threw a party for him but by oversight, Peierls, then head of the Los Alamos British Mission, did not receive an invitation. The next day Oppenheimer sought out Peierls to apologize. "This is terrible," he told Peierls, "but there is an element of comfort in the situation. It might have happened with Edward Teller."[15]

Even when it became clear that a fusion weapon could not be perfected before the probable end of hostilities, Teller was allowed to continue his research. Moreover, he received considerable support in this endeavor from numerous British Mission personnel.

Egon Bretscher, a Swiss émigré to Cambridge, worked closely with Teller in this program. An outgoing but high-strung man, who often suffered from migraine headaches, Bretscher sometimes manifested the air of a complainer. (At Cambridge he had once reputedly said: "I don't know what's wrong with me today. I feel fine."[16]) His reputation did not diminish in Los Alamos, for he complained vigorously during his initial months that he had nothing to do. After several long hikes with his old friends, Hans Bethe and Enrico Fermi, Bretscher finally found support for his ideas. Oppenheimer then placed him in charge of a special group F-3, Super Experimentation, that worked with Edward Teller on the question of fusion. Bretscher had enough support from Oppenheimer to request that two junior people from Cambridge – French and Poole – join him.[17] This team did numerous preliminary measurements regarding the thermonuclear bomb.

The first task of F-3 was to make a low voltage accelerator, a Cockcroft-Walton machine up to about a hundred kilovolts acceleration. This was then used for deuterium-deuterium and deuterium-tritium cross-section measurements. It took about a year to get the machine working in all dimensions. Eventually, Hanford sent them one cubic centimeter of tritium gas, and this small amount provided the raw materials from which they had to do most of their cross-section measurements. (A second cubic centimeter was provided later.) It was the realization that Hanford simply could not produce enough tritium that dimmed hopes for a fusion weapon by 1945. Only Teller refused to give up.

The work of the F-3 group appears in few Los Alamos documents, largely because of the secrecy surrounding the hydrogen bomb. Even though it was over a year before the plutonium bomb proved viable, Bretscher's team devoted its research interests elsewhere. French and several of the others did not even journey to Trinity Site to watch the July 16 test. As French recalled in 1987, "never at any time did I have anything to do with the fission bomb once I went to Los Alamos."[18]

The fusion team's experiments revealed a huge cross-section for the tritium-deuterium reaction. In fact, when they first substituted tritium for deuterium as their ion source, after testing out the system with the deuterium-deuterium reaction, the rain of pulses on the oscilloscope made them at first suspect electrical breakdowns in their detectors. This discovery pointed to the sobering potential for some form of successful fusion weapon in the future. Over forty years later, French still viewed these experiments as the most dramatic moment of his scientific career.[19] Poole and French later wrote their doctoral dissertations on this subject, and James Tuck also wrote one of the first papers on the possibility of fusion power. But the development of an actual fusion weapon had to wait until the Ulam-Teller theory of the early 1950s. This theory, still one of the nation's most closely guarded scientific secrets, suggested a completely different approach to a hydrogen bomb from that which had been contemplated during the war years.

As part of their contribution to the Manhattan Project, Churchill's government tried to select scientists who had specialized knowledge and skills that the Americans lacked. One of these areas lay with the effects of bombing and blast waves. Hence, virtually every reminiscence lauds the efforts of G.I. Taylor and William G. (now Lord) Penney.

A scientific polymath who was later described as "the personification of the peculiarly British tradition of applied mathematics," Taylor had worked on a wide variety of defense problems during the early war years.[20] These involved the effect of underwater explosions on various structures, the dispersal of fog from runways by lines of gas burners, rocket shapes, and shaped explosive charges for piercing armor plate.

But it was Taylor's knowledge of spherically expanding blast waves that made him so valuable to the American scientists. He was the author of the "Taylor instability" theory, which suggested that when a light material pushes against a heavier one, the interface between them will be unstable. His advice proved of considerable help to the T Division, and his ideas stimulated much discussion about the ball of fire phenomenon. He correctly predicted many of the atmospheric phenomena at Trinity, including the mushroom cloud, the height of the explosion, and the effect of the wind on the distribution of particles.[21] The Americans wanted Taylor to be present at Trinity so much that, as Chadwick wrote, "Anything short of kidnapping would be justified."[22]

William Penney had devoted much of his early wartime service to the effects of German bombing on England. He too was highly skilled in the calculation of blast and shock waves and estimation of bomb damage. Unlike Taylor, Penney became a permanent member of Los Alamos. He also served as a special assistant to Kenneth T. Bainbridge, the Harvard physicist who was overall director of Trinity Site. When it appeared that Penney might return to his post in Imperial College before the war's end, Groves put in a special request that he stay.[23] His calculations eventually proved vital in determining the height of detonation for the two Japanese bombs.

In addition to the theoretical contributions, the British Mission team also contributed a number of practical inventions that accelerated the timescale of the Manhattan Project. Of these, the most important came from Otto Frisch and James Tuck.

Otto Frisch moved easily between the theoretical and experimental worlds. A self-confessed "non organizer," Frisch spent much of his time at Los Alamos as an ombudsman, moving from problem to problem. Perhaps the most important problem he worked on was to determine exactly how much enriched uranium would be needed for the U^{235} bomb. When the first shipments of uranium arrived from Oak Ridge, Frisch suggested a plan. He would assemble a block of metal hydride, similar to a bomb, but would leave a hole in the center. He would then place another piece of uranium on a track and drop it through the hole. For an instant, the two pieces would create the conditions necessary for an explosion. When he first mentioned this scheme, Richard Feynman compared it to "tickling the tail of a sleeping dragon." Thus, it became known as the "dragon experiment."[24]

Frisch took over the isolated Omega site in a Los Alamos canyon for his experiment. For several weeks, Frisch's Critical Assembly Group was one of the few teams to work with nuclear materials on a day-to-day basis. The test was completed on January 18, 1945, and it worked perfectly. It was, as Frisch recalled "as near as we could possibly go towards starting an atomic explosion without actually being blown up . . ."[25] The results of this experiment obviated the need to field test the uranium bomb. There was no question that the "gun assembly" would detonate as anticipated. Necessary though it might have been at the time, the dragon experiment was also foolhardy. Shortly after the end of the war, scientists Harry Daghlian and Louis Slotin were both killed in performing similar tests.

The other major British Mission experimentalist was James Tuck. Described as "a remarkable combination of social *naïveté* and technical astuteness," Tuck began his career as a scientific assistant to Professor F.A. Lindemann at the Clarendon in Oxford.[26] Lindemann brought Tuck with him when he went to Whitehall as Churchill's science advisor. Tuck, however, disliked the desk work and soon joined MD1 ("Winston Churchill's toyshop") which worked on problems of unconventional weaponry. There he proved quite valuable in elucidating the mechanism of the high explosive Munroe (shaped charge) effect, used in anti-tank weapons for armor piercing; he also did work on magnetic mines, and numerous flash X-ray experiments.[27]

When Tuck arrived in Los Alamos in 1943, he discovered that the news of his experiments had preceded him. Several of these studies indicated that optical principles could be used to shape detonation waves. Tuck insisted that uniform shock waves could be created by "shaping" explosive charges so that they would produce the desired compression wave. The concept was elaborated on by mathematical physicists, especially the Hungarian genius, John von Neumann. Everyone wrestled with the implosion issue, even General Groves. He devised a scheme – taking a shell and blowing it out – that the scientists knew would not work; still, "The General," as the scheme was called, had to be tested. Tuck worked especially hard on the explosive lenses of the implosion project and finally, in late 1944, Oppenheimer reported his experiment to the Coordinating Council, saying "for the first time, solid matter has been visibly compressed. Lens implosion was going to work."[28]

In addition to the lens system, Tuck designed the initiator (urchin) for the Trinity bomb. As he later noted, "The lens and the urchin came to be components in the Trinity shot, and vital to its functions, and, for that matter, to the functioning of all subsequent weapons for many years."[29] The American scientist who worked most closely with him on this was Seth Neddermeyer, the inventor of the implosion method of detonation. Most accounts usually credit just Neddermeyer,[30] but American physicist Bruno Rossi later told British scientist R.V. Jones that without Tuck's contributions to the fusing mechanisms, the plutonium bomb could not have been exploded in August of 1945.[31]

In addition to Frisch and Tuck, the successful Trinity site detonation in July relied heavily on several other British contributions. Ernest Titterton served as senior member of the Timing Group. P.B. Moon worked as Bainbridge's assistant, and wrote three important pre-Trinity reports on the possible consequences of the first nuclear blast. He also anticipated several of the later experiments. Moon's duty was to establish ways of measuring fall-out radiation levels from instruments that he helped produce. Such measurements of radiation levels were then given to the Health Safety people who utilized them to monitor safety levels.[31] Marley, Sheard, and Littler also assisted in blast measurements at Trinity.[32]

The extensive photographic record of the Trinity explosion owes much to the British. Gregory Marley brought with him to the States two special cameras, weighing 322 and 135 pounds, respectively. These played an important role in improving American high-speed photography, for the British cameras were initially more sophisticated than the Americans'. Marley also invented a camera that took 100,000 pictures a second, and this proved invaluable in both implosion research and the recording of the blast wave. The extensive blast measurements from Trinity also relied on British expertise. Littler and Sheard worked closely with John Manley in placing Piezo electric gages and "bursting disks" all through the region. Their analysis of energy yield of Trinity helped determine the height of detonation for the Japanese bombs.[33]

Most of the British team journeyed to Trinity Site to observe the test on July 16. Those at the Base Camp (ten miles from Ground Zero) were told to lie on their stomachs and look at the flash of light only through a piece of welder's glass. D.J. Littler lost the opportunity to see the first few seconds of the explosion because he was so distracted by the intense burning sensation on the back of his neck. Stationed at Campania Hill, a small knoll about 25 miles away, Chadwick later described the blast as: "a great blinding light [which] lit up the sky and earth, as if God Himself appeared among us . . . There came the report of the explosion, sudden and sharp as if the skies had cracked . . . a vision from the Book of Revelation." Otto Frisch's description of Trinity (see Appendix IV) has become a minor classic. It was, remarked Frisch, "as if somebody had turned the sun on with a switch." Yet it was probably James Tuck's comment that summed up the feelings of the majority. Said Tuck: "What have we done?"[34]

As American scientist Philip Morrison recalled, the successful Trinity Site test probably served as the highlight of the Los Alamos experience for many scientists. Up until then, he said "the whole thing was an intellectual game – to defeat the enemy through an intellectual accomplishment."[35] In a strange sense, Hiroshima proved almost an anticlimax.

Some of the other British Mission contributions to Los Alamos include: William G. Penney and Geoffrey I. Taylor wrote crucial reports on blast effects. James Hughes helped construct the Cockcroft-Walton machine. Group leader Rudolf Peierls encouraged Robert Christy in his invention of a solid implosion device, usually termed "the Christy gadget." George Placzek was called on for his expertise in neutron diffusion theory, a knowledge that had not been reduced to writing at that time. Donald G. Marshall did valuable field testing to help determine the critical size and shape of the nuclear cores. Peierls and Frisch later wrote volumes in the *Los Alamos Technical Series* (originally termed the *Los Alamos Encyclopedia*); Marley, who advised the implosion groups, also wrote several chapters. British Mission personnel were the sole inventors of nine items – Bohr (1), Frisch (5), Poole (1) and Titterton (2) – and joint inventors of seven others.[36] The list of other contributions could easily be extended. Ralph Carlisle Smith's 1949 summation of their accomplishments ran to five pages. (See Appendix III.)

The presence of British nationals working alongside American personnel on sophisticated technical and scientific problems raised a variety of questions. Few were more complicated than those relating to patent matters. For US civilian and military personnel, the patent question remained straightforward. In return for payment of one dollar, all patents became the property of the US government. The British operated under a similar system but there were a number of subtle differences. The problem of joint American-British patents, of course, was unique.[37]

Consequently in June, 1944, J. Robert Oppenheimer called Army Captain Ralph Carlisle Smith into his office. A chemical engineering graduate from Rensselaer Institute of Technology, Smith had been working as a patent attorney before he joined the military. In 1943, he had been sent to Los Alamos as a patent representative for the OSRD. In June of 1944, Oppenheimer assigned him the task of monitoring all American and British patent applications.

As Smith recalled his instructions, Oppenheimer told him: "You have a very tough job. You know that you are to protect everything patentable. You have complete access to everything in the laboratory, the reports, anything and all, and any facility. You have one instruction from me: do your job, but don't interfere with the work."[38] Overall, Smith handled about 500 patent applications during the war years.[39]

Because of this assignment, Smith found himself sitting in on literally hundreds of meetings and conferences. He carefully jotted down which British Mission personnel attended which meetings and what items were discussed there. From these notes he compiled a monthly report of British Mission activities at Los Alamos. At the direction of Oppenheimer, he sent these on to Groves in Washington.

Since Smith's monthly reports also included British Mission arrivals and departures from Los Alamos, Army intelligence also began to draw upon them. They served as a convenient means of monitoring the activities of all British Mission members. Smith protested that his duties involved patent protection, not security, but that objection was ignored. None of the British Mission were aware of Smith's assignment at the time.[40]

Army security remained very tight for all Manhattan Project sites. Groves maintained about a 250-person security force nationwide, and many of these were stationed in New Mexico. Security men tailed the Los Alamos scientists whenever they left the site. Teller, Oppenheimer, Bethe, Fermi and others all had their "shadows."

Army intelligence (G-2) also covered Santa Fe. The bartender at the La Fonda Hotel on the plaza worked for G-2, as did several other hotel employees. The same was true for both the Hilton Hotel and the Alvarado Hotel in Albuquerque. Judy Florence, who had grown up in a New York City radical family and happened to be at the University of New Mexico studying anthropology, was enlisted for this duty. Because of her background – her father had been a well-known radical – she had met many of the old left-wingers and knew them personally. Thus, she worked for intelligence as a waitress at the Hilton. Her instructions were to watch for certain people. If any Los Alamos scientists had too much to drink, she was to call a cab to take them home. The cab drivers worked for security, too.

Los Alamos never devised any official "cover story" to explain the presence of so many British accents to local residents. So, each Mission member had to come up with his own disguise. Donald Marshall and James

Hughes elected to become "clerks" from the British Embassy in Washington on vacation to the Southwest. Once they met some University of New Mexico students in Albuquerque who insisted on trying to show them the town of Santa Fe. "We had to duck out of that pretty smartly," Marshall recalled.[41]

The American scientists were not watched on-site, but many of them were followed off-site. The reverse was true for the British. Smith covered all their on-site activities, but nobody followed them once they left the Hill. This anomaly would later bear strange fruit in the espionage of Klaus Fuchs.

While the British Mission members were frequently called on for theoretical and technical expertise, they were not similarly consulted on the political or military aspects of the atomic weapon they were creating. This decision remained in the hands of Washington and Whitehall.

When William Penney arrived in Los Alamos, for example, Oppenheimer called him in and told him that the politicians in Washington would have the final say regarding use of the atomic weapons. The Los Alamos British Mission was never consulted, as such, on this matter. While some British scientists stationed elsewhere – Sir John Anderson in Canada for example – expressed anxiety about the proposed use of the weapon, there was never a concerted "British debate." Since their scientists were scattered across the US, that simply was not feasible. William Penney played an important role in target selection, and Chadwick in Washington stayed in daily contact with Groves. But everyone assumed that the British would "go along" with any decision made by the Americans. In a July 4, 1945, meeting in Washington, the British gave their consent "in principle" to the use of the weapon. After that, as Winston Churchill phrased it, "the final decision now lay in the main with President Truman."[42]

The relative influence of the two nations may be illustrated in the rivalry over which plane would be utilized to drop the weapon. The British Lancaster bomber had a longer range than the American B-29, and the Lancaster needed fewer special modifications than the Boeing aircraft. Initially Los Alamos had planned to modify a Lancaster for this purpose but gradually they changed their minds. In 1943, at Oak Ridge, Lord Cherwell and James Chadwick had engaged in a vigorous debate with Groves and Conant on this issue. The Americans maintained that the slow and rather clumsy Lancaster did not possess the range to operate in the vast Pacific theater. But the real reason was their determination not to use a British aircraft for this mission. As General H.H. Arnold, Chief of the Army Air Force, stated: "There was no way they were going to deliver the American atomic bomb in a British plane."[43]

Neither Congress nor Parliament knew about the joint weapons project; nor did the Cabinets of either government. Only the top figures of the Anglo-American military were brought into the secret. The statesmen who were planning the infant United Nations were also kept in ignorance. While the British and American elite policymakers met regularly to discuss the matter, the Americans always took the lead. The ultimate "decisions" about atomic weapons were made by a precious few.

Moreover, the Americans did not keep the British leadership at 10 Downing Street well informed about Los Alamos. In 1953, when Churchill was examining documents for Volume VI of his *Memoirs*, he discovered that the Americans had not notified him that the Trinity test was forthcoming. He was decidedly annoyed about this. He confessed that had he known that the success of Tube Alloys was so imminent, he might have postponed the date of the General Election that unseated him.[44]

The Americans also did not share the immediate news of the Trinity test with Churchill. After the test results were cabled to Washington on July 16, 1945, British Ambassador Lord Halifax had a lengthy conference with Groves on the issue. For unknown reasons, Halifax decided not to send a full report to Churchill at Potsdam at that time. Thus, Groves' report to Truman provided the only official notification of the Trinity results. Earlier, Churchill had asked Truman to let him know if it were "a plop or a flop" and Truman cabled him only "It's a plop."[45] Churchill was not given any details until July 22, when Stimson briefed him in full. Afterwards Churchill went to see Truman and they talked for an hour. By then, however, Churchill had become a lame-duck Prime Minister, and the British role in the Pacific campaign diminished even more. In a strange sense, both the atomic bombs and ending the war against Japan had emerged as America's affair.[46]

Neither Churchill nor Truman ever agonized over their 1945 decision to drop the atomic bombs on Japan. Churchill acknowledged that history might, indeed, question his judgment in this matter, but he remained firm in his conviction. "I may even be asked by my Maker why I used it," he wrote later, "but I shall defend myself vigorously and shall say, 'Why did you release this knowledge to us when mankind was raging in furious battles.'"[47]

Truman remained equally adamant. He told a reporter that he slept soundly the night after he had given his order. One looks in vain for any second guessing in his numerous comments on this decision. "It was a question of saving hundreds of thousands of American lives," he wrote later. "I could not worry about what history would say about my personal morality." As he remarked to Congress, the bomb did not win the war, but it certainly shortened it.[48]

An integral part of the Truman/Churchill "decision," and the widespread public acceptance of it, may be seen in the massive Allied air raids that preceded it. From 1944 forward, both British and American air forces shifted from "precision" bombing (aiming for specific military targets) to "saturation" bombing (the obliteration of whole cities to destroy morale as well as industrial sites). This shift in tactics meant that the air raids over Hamburg, Berlin and Dresden all produced firestorms of incredible devastation.[49]

In the fall of 1944, the Army Air Force began dropping the newly invented napalm over the Japanese islands. The collapse of the Japanese Air Force made the fire bombing of Tokyo in 1945 a relatively easy task. The military planners had rationalized such incendiary attacks by stating that since the Japanese government had mobilized the Japanese people, the entire civilian population had become a legitimate military target. Dramatic newspaper headlines on the bombing raids conditioned the home population. By 1945, the American and British public approved of virtually any type of bombing.[50] "Without this background," Rudolf Peierls later observed, "the atomic bomb raids on Japan might not have taken place."[51]

Wartime security measures stymied any "debate" on the use of the atomic weapons, either on a cabinet, popular, or scientific level. Only isolated voices cautioned against use. The most famous of these voices – the James Franck Report from the University of Chicago and the petition from maverick scientist Leo Szilard – never received a thorough hearing. *New York Times* reporter William L. Lawrence later remarked that at the time he never heard a word from the Los Alamos scientists about the moral implications of the bomb.[52] The daily news Bulletins at Los Alamos regularly stressed that the importance of the Project would not end with the collapse of Germany. One May, 1945, directive from Oppenheimer and Undersecretary of War Robert Patterson urged the workers to continue until Japan was "completely crushed." Other bulletins called for "quicker taps for Tokyo."[53] The announcement of V-E day brought only brief respite: a large party on the Hill where Frisch, Peierls, Bethe and others astonished their colleagues by singing old German student songs. But the fall of Germany had no real impact on the race to ready the weapons for combat use. Moreover, there is no evidence that the British Mission at Los Alamos expressed any opinion either way at the time. Even if they had, it is doubtful that it would have been heard in Washington.

Journalist Peter Wyden's 1984 study *Day One: Before Hiroshima and After* (and the 1989 television drama, "Day One," based on the book) have implied that the voices opposing the dropping of the bomb were roughly equal to those who favored using it. The position may be artistically viable, as balance makes for better drama. It certainly may be morally feasible.

But, unfortunately, it is not true to the historical record. The scientists who opposed the combat use of the weapon before August, 1945, were a small, out-of-power minority. The Manhattan Project was no democracy. Power flowed from the top down, and the top held a single opinion. Arthur Compton recalled that everyone took it for granted that the bombs would be used in Europe if they were ready in time. Churchill remembered the same emotion. "[There] never was a moment's discussion as to whether the atomic bomb should be used or not," he wrote in 1953. "The decision whether or not to use the atomic bomb to compel the surrender of Japan was never even an issue. There was unanimous, automatic, unquestioned agreement around our table."[54] On June 1, 1945, the Interim Committee delivered a unanimous report to President Truman. It recommended that the bomb be utilized on Japan as soon as possible; and that it be dropped without prior warning.[55]

The rise of the doubts over the morality of Hiroshima and Nagasaki came afterwards. D.J. Littler recalled that several American physicists greeted William Penney's pre-Hiroshima description of probable Japanese casualties with "a certain amount of enthusiasm." The British scientists, however, most of whom had lived through German bombing, remained a bit more circumspect. A.P. French did not recall any moral qualms at the time of the successful Trinity test, but he did remember being shaken when he first heard the news of Hiroshima. Donald G. Marshall, however, recalled that the dominant emotion after the Hiroshima announcement was primarily that of "triumph and relief." Groups began to arrange for various celebrations and Marshall formed part of one team sent to Santa Fe to purchase all the alcohol they could lay their hands on. "We were jubilant," he said.[56]

Tony Skyrme noted that, in general, the British Mission personnel felt the bomb was a good thing. "We thought that once the bomb had been demonstrated, the world would see how terrible this weapon was and come to its senses," Skyrme observed.[57] Alice Kimball Smith also stated that most of the people at Los Alamos believed that "peaceful applications of atomic energy to power, medicine, agriculture, and industry would in the long run far outweigh destruction."[58] Chadwick certainly felt that way. He always emphasized the medical applications of his research in nuclear matters. Later he confessed that until the very end, he had often hoped that the bomb would not work.[59]

Sir Henry Tizard, however, felt otherwise. In 1946, he declared in a major address that the world was fortunate that both the atomic bomb and the long-range rocket had been publicly demonstrated before the end of the conflict. "They opened wide the eyes of the people of the world," Tizard said, who might otherwise have allowed the weapons to be developed in secret.[60] For those who were combat soldiers at the time, such as William T. Hagan, Clarence Critzman, and Paul Fussell, there was no question that President Truman made the proper decision.[61] Soldiers wearing the Manhattan Project arm patch were often sought out and thanked by other GIs, especially those who had been poised to invade Japan.[62]

In his autobiography, *Bird of Passage*, Peierls wrestled long and hard with this question. He concluded that it was not so much the *scale* of destruction but the ease and suddenness that made the atomic bomb such a force. One plane could now do the work of thousands. He deemed the risk of not having the bomb – when Germany could possibly be working on it – as one too large to ignore. A "strike" by the British scientists was not possible. Peierls, however, concluded that the scientists probably should have insisted on more dialogue with the military and political leaders.[63] D.J. Littler accepted the standard Allied arguments for the use of the bomb on Hiroshima but expressed grave doubts about the need for the second weapon. P.B. Moon declined to reflect on the moral issue, arguing that such reflections are not common among physicists, nor among people in general.[64] But as Peggy Titterton noted, it was chiefly fear of failure, not concern over casualties, that remained the dominant emotion at Los Alamos.[65]

In his autobiography *What Little I Remember*, Otto Frisch also recalled that he had no initial misgivings about the project. The contents of the natural world could be read as easily by the German scientists as by the Allies. But he joined Peierls in wishing that the first bomb could have been dropped on an unoccupied Japanese island and not a crowded city. In Frisch's case, however, action (more precisely, inaction) spoke louder than the words of his autobiography. Frisch's friend, J.G. Rushbrooke, observed that never after 1945 did he hear Frisch comment on politics. Frisch's involvement with the Tube Alloys/Manhattan Project left him with a permanent distaste for the contemporary political world.[66]

3 The British Mission at Los Alamos: The Social Dimension

Like their American counterparts, the British Mission personnel had no idea what to expect when they were assigned to Los Alamos. Most of them boarded the California Limited in Chicago and disembarked at Lamy, New Mexico (the closest railway station to Los Alamos), about ten miles from Santa Fe. They then were met by car, usually driven by a WAC, and taken to 109 East Palace Avenue in Santa Fe. There they were introduced to Dorothy McKibben, who headed the Santa Fe office, and received their passes to Los Alamos. All were astonished by the adobe architecture, the high desert vegetation, and the expansive horizon of the American Southwest. One scientist recalled wartime Santa Fe as "something out of a western set."[1]

Since the Hill community grew steadily all through the war period, Los Alamos remained a sea of construction. Oppenheimer instructed the Army bulldozer drivers to utilize the natural contours of the land as much as possible, but the streets and houses retained a jerry-built appearance for the duration. No one expected Los Alamos to survive the war. Consequently, everything had a temporary feel to it.

The houses, which were all wood and painted green, were scattered without pattern on streets with no names. There were no sidewalks. One retained his or her orientation by a single landmark – the large water tower that was visible from all sections of the mesa. When Eugenia Peierls first arrived, she expressed dismay that the lovely mountain landscape had been so spoiled by the erection of army huts. But when the Peierls returned in the mid-1960s to view the small, nondescript row houses of a now-permanent Los Alamos, they almost wished for the huts back.[2]

Of the British Mission, only the Chadwicks and Tittertons lived in homes that dated from the ranch school days. Even these necessitated much remodeling. When Chadwick decided to bring his family, Groves authorized the renovation of a large log cabin, at the rear of the main building, Fuller Lodge. This rebuilt mechanic's home became one of the most desired houses on the Hill. The Tittertons lived in a converted log cabin guest cottage that is now a part of the Los Alamos Historical Society Museum.

The old ranch school buildings usually contained all the amenities and, consequently, were dubbed "bathtub row" by the others. But the majority of the British scientists found life on the primitive side. A.P. French recalled his Los Alamos dorm room as cramped and bare, especially when compared to his relatively luxurious college rooms at Cambridge. Some bachelors, such as Joseph Rotblat and Klaus Fuchs, were housed in the Big House (no longer extant), while others were scattered about in Single Men's Dormitories. There they groused about the high temperatures at which Americans kept their buildings. The single men usually ate their meals at the nearby East Mess or at the newly built cafeteria next to Dorm 8. The fortunate ones dined at Fuller Lodge. The Lodge gave such good value for the money that few could turn it down, in spite of a six-month waiting list. Families also dined in Fuller Lodge, especially on Fridays when the chefs served up steaks for the unheard of bargain of $1.00 a meal. Groves felt good food was essential to morale and he never skimped along those lines.

Married couples, such as the Peierls and Bretschers, were assigned to one of the hastily built four-plex Sundt homes. The Peierls found the move to America especially rewarding, for they could now rejoin their two children, who had been evacuated to Canada after the Germans had begun bombing London. All inhabitants of the Sundt homes wrestled with an inadequate heating system, and a temperamental wood burning cook stove termed "Black Beauty." The contrast between the rugged living conditions and the overall calibre of the inhabitants proved startling. "Out of the shacks came suddenly great numbers of men in evening dress and ladies as elegantly attired, sort of walking through the mud," mathematician Stanislaw Ulam later recalled. "It was the most incongruous thing I could imagine."[3]

The saving grace came from the physical environment. Every Los Alamos house offered a view of the Jemez mountains or the spectacular canyons that cut sharply through the mesa. A.P. French fell in love with the surroundings immediately. Accustomed to the clouds and small sky of England, he and Poole were perpetually entranced by the blue sky and vast southwestern horizon. During the winter months, they often rose early so they could watch the sun break over the Sangre de Cristo Mountains to the east. British-born metallurgist Cyril S. Smith considered the spectacular environment of Los Alamos as much a part of the Tube Alloys/Manhattan Project as the leadership of Oppenheimer or Groves. Jim Tuck and Donald Marshall both saw much of the surrounding country from horseback, as they became devoted "horse riders" (a pastime limited to the wealthy back home). The Laboratory closed down almost completely on Sundays, and many of the British Mission spent the day hiking through the countryside.

These Sunday hikes made it possible for the scientists to maintain the intense level of work for the other six days.[4] Without the release provided by the parties and fresh air, teacher Jean Bacher later noted, "we would have gone mad."[5]

The scientists from foreign backgrounds carried their traditional love of hiking to the Jemez and Sangre de Cristo mountains of New Mexico. Fuchs, Bohr, Bretscher, Smith, Peierls, Fermi, Tuck and even Teller, who had an artificial foot, all enjoyed vigorous hikes. Several members of the Mission also developed an ethnographic interest in the nearby Spanish villages and Indian cultures, and returned home with Navajo rugs, Pueblo pottery, and various types of turquoise jewelry. Many of them attended the famous Indian festival at Gallup, New Mexico, or the Winter Solstice Shalako celebration at Zuni Pueblo. The Anasazi ruins at nearby Bandelier National Monument became their private playground.

Perhaps the master hiker was Egon Bretscher. Bretscher had grown up exploring the mountains of his native Switzerland (he once actually saved Felix Bloch's life in a climbing incident). At Los Alamos, he carried this passion to the Jemez Mountains. Once Bretscher left through the main gate, signing out as was customary. The Army guard glanced at his reason for leaving and added three exclamation marks after Bretscher's "walking!!!"[6]

Naturally, nearby Santa Feans wondered about all the activity on the Hill. Their guesses about the purpose of Los Alamos ranged widely, but they included: the manufacture of rockets, jet propulsion, "death rays," atomic energy, some type of bomb, or chemical warfare. The consensus lay with a type of poison gas. Dorothy McKibben, however, used her considerable prestige to keep such inquiries to a minimum.

The isolation of Los Alamos reflected Groves' passion for security. Travel from the Hill was restricted, and the scientists were not even supposed to visit their Santa Fe friends on a regular basis. When Otto Frisch played piano over the local radio station, KRS, he was introduced simply as "our pianist," as even the mention of his name was seen as giving the game away. Other scientists, such as Arthur Compton, Niels Bohr, and Aage Bohr were all given pseudonyms. Enrico Fermi annoyed the authorities by refusing to take his ("Henry Farmer") seriously.

All outgoing mail from the Hill was censored, and all incoming mail went first to Box 1663, Santa Fe, New Mexico. All incoming British mail was initially sent to a Washington, DC mail drop and then forwarded to Los Alamos under cover of British embassy mail. Titles such as "Professor" or "Doctor" were forbidden, as well as terms like "physicist" or "chemist." The Army specifically requested British officials that letters to Chadwick delete both the "Sir" and "Dr."[7] The absurdity of this seemed obvious to many. When physicist Sam Allison once asked Enrico Fermi who nine-teenth-century Archbishop Jean Baptist Lamy was, Fermi silenced him with: "Mrs McKibben would suggest we call him 'Mr Lamy.'"[8]

The secrecy and isolation of wartime Los Alamos meant that the community was thrown back upon itself. Army buses traversed the road to Santa Fe once a week for shopping trips, rarely exceeding the 35 m.p.h. speed limit. Some of the Americans had automobiles, but of the British Mission, only Fuchs and Peierls could afford them. Side trips to installations on the mesas were always made in jeeps borrowed from the Army pool. The Americans were usually eager to include the British in their side ventures, however. The Mesa was very small, Peggy Titterton recalled, and you had to know and get along with everybody.[9]

Felix Bloch once warned Hanni Bretscher that Los Alamos was a very strange place. A person could never plan to spend an evening reading a book or writing a letter because somebody was certain to drop in.[10] Otto Frisch recalled that he could walk into virtually any home on the Hill and soon find himself engaged in a stimulating discussion over issues of art, music, or politics.[11] The weekend parties reflected this need for self entertainment. The guests often played parlor games, such as "Twenty Questions" which Fermi delighted in, or charades, which Fuchs enjoyed.[12] Tony Skyrme, whom one member recalled as "very English," "very eccentric" and "mad as a hatter," especially enjoyed playing difficult parlor games.[13]

The Los Alamos community shared the esprit of a battle unit. The political discussions at dinner and afterwards frequently became intense. Here the scientists and their families debated the probable political fate of postwar Europe. It was at Los Alamos that the concept of a divided Germany received its first semi-public analysis. The social life at Los Alamos was extensive and very good, Joseph Rotblat recalled, "They were very stimulating discussions."[14] As Rudolf Peierls noted, however, "there was not much else to do."[15]

Many of these social gatherings revolved around music. A surprising number of the British scientists or their wives – Ernest Titterton, Egon Bretscher, Otto Frisch and Winifred Moon – were accomplished musicians. Frisch had only to hear a piece once and he could play it perfectly on the piano. He often played a half hour of piano at noon over KRS. In July of 1944, Frisch gave a formal benefit concert to help raise funds for USO projects. Everyone acknowledged his ability as near concert level quality.

E.W. Titterton also played classical music, but he preferred jazz and popular tunes. Consequently, he was much in demand at the social gatherings. Since the Tittertons lived adjacent to Fuller Lodge, site of the lone grand piano, he could easily disappear to play by himself. Everyone enjoyed Titterton's keyboard skills and one fan even bought him sheet music for several Strauss waltzes. Physicist John Manley, usually described as Oppenheimer's right-hand man at Los Alamos, once arranged a series of parties on the principle that every guest be skilled either in music or in Chinese cookery. On these occasions, the hosts chopped vegetables to a backdrop of a violin, cello, and piano trio.[16] Edward Teller's piano artistry has become legendary, largely because he insisted on playing at all hours of the night so as to relieve tension. There was always music around, John Manley recalled.[17]

Because they had to entertain themselves, the Los Alamos community proved very ingenious. They skied, hiked, and played baseball and basketball. In addition, they established a public library to fend off the long winter evenings and frequently held book discussions. They also put on a homegrown circus, complete with acrobatics for children. The local thespians staged several dramatic performances. These included "Arsenic and Old Lace," "Hay Fever," "Right About Face," "You Can't Take It with You," "Dangerous Corner," "Outward Bound," "Claudia," "Kind Lady" and "The Male Animal."[18] At Christmas, a choral group staged Handel's "Messiah." Moreover, the Army showed one new feature film every night of the week. Although they remained isolated, the Los Alamites thus had an opportunity to sample everything that wartime Hollywood produced for "the outside world."[19]

The most popular social gatherings, however, were the dance parties. These came in many versions. The Square Dance Club met every other Saturday from the beginning of the project (it is still in operation). Through it, many British Mission members learned to square dance. In turn, the British and Europeans impressed the Americans with their skill at ballroom dancing. Klaus Fuchs, for one, was much in demand on such occasions. As Carson Mark described him, Fuchs "was an excellent man of the waltz."[20] Alice Kimball Smith recalled one incident at the dances. A tall woman, Smith felt embarrassed to dance with shorter men, and once declined an invitation to waltz with Fuchs. Within a few moments, however, she found herself on the floor with Edward Teller, who was much shorter than Fuchs. Over forty years later, Smith still recalled the incident with embarrassment.[21] Ruth Marshak always maintained that the intense level of social activity was essential to the health of Los Alamos. The scientists and their wives, she said, "not only worked inhuman hours to perfect the bomb; they also had energy to dissipate on skiing and horseback riding, mountain climbing and folk dancing, and gay parties which lasted until dawn."[22]

A number of British Mission wives – Hanni Bretscher, Peggy Titterton, Winifred Moon, Elsa Placzek, Genia Peierls, and Elsie Tuck – accompanied their husbands to Los Alamos. Most arrived with their husbands, but James Tuck had to agitate the authorities for some time before his wife, Elsie, was allowed to join him. "I shall blow up this place proper, you know, if my little wife doesn't get oveh heah," he once said. "She's a tiny little one, no bigger than a minute, won't take up much war space."[23] Eventually, a frightened Elsie Tuck boarded a ship for New York City, where she was met by a complete stranger. The man placed her on a train to Chicago (she did not know her destination) and three days later she was finally reunited with her husband.[24]

Laboratory officials tried to utilize the wives' skills whenever possible. Winifred Moon worked as a secretary, and Peggy Titterton, trained as a laboratory technician, worked several months in the Tech area. Other women, such as Jane Wilson, Alice Smith and Ruth Marshak, taught in the Los Alamos school system. On the other hand, physicist Genia Peierls and mathematician Hanni Bretscher did not work in the schools or laboratory; instead, they devoted their hours to household duties. So, too, did Kathleen Mark. All had small children and household help was hard to come by. "The men didn't have time to do anything," one British Mission wife recalled, "anything at all."[25]

The wives who did not work outside the home, however, entered whole-heartedly into what was termed "community affairs" or "Mesa Business." Mesa Business soon assumed serious proportions. Genia Peierls, for example, devoted much of her time and energy to this aspect of life. Boisterous and opinionated, Genia Peierls also manifested a genuine interest in other people and their predicaments. Recognized as the life of the party, she was soon viewed as a Los Alamos "character." When she first heard about the success of D-Day, for example, she danced on the tables. On another occasion, Genia Peierls decided that the level of science teaching at Los Alamos High School, where their daughter Gabi attended, was not acceptable. Consequently, she organized a program whereby several laboratory scientists visited the school to give lectures on their specialties. This proved eminently successful.[26] A third Genia Peierls story involved the library. Annoyed that so many books were overdrawn, she decided to remedy the situation personally. So, she borrowed Rose Bethe's baby-buggy, went from house to house (nobody locked their doors) and simply borrowed back all the library books.[27]

Genia Peierls played a prominent role in Los Alamos social life. Her broad frame, resonant voice, and distinctive accent (she was fluent in English but usually skipped all articles, e.g. "We are Peierls") often made her the center of conversation. Rumors flew behind her back that she had once been in the Russian Army, with a rank that varied from private to captain. Her flamboyance provided a marked contrast to her quiet, soft-spoken husband. Genia Peierls once boasted that all the worthwhile physicists in the British Isles had spent at least one evening in their home. She also claimed that all international visitors of note had stopped there, too.[28] Nobody could ever forget Genia Peierls, recalled John Manley.[29]

James Tuck assumed a similar high profile in the British Mission entourage. Viewed by the Americans as a "typical comic Englishman," Tuck's demeanor, pronounced accent, handsome visage ("more Oxford than Oxford") and sense of humor endeared him to everyone. His reputation continued to grow when he moved back to Los Alamos in 1950.

Tuck's decision to relocate on the Hill was directly related to the entrenched social class system that still infused the postwar British universities. In 1946, Tuck returned to Oxford but a friend had warned him that his wartime successes would probably not be reflected in his position in the Clarendon Laboratory hierarchy. The same situation could even be found at Harwell. For example, during a food shortage one aristocratic staff member there tried to woo those in power with hampers of home grown vegetables and fruit. Tuck once met this person, fruit basket in hand, in front of John Cockcroft's Harwell office. The man was trying for a scientific grant, but Cockcroft dismissed him abruptly with: "It takes papers here, not pears."

In 1949, Edward Teller convinced Tuck to shed these frustrations and return to the States. After a year at the University of Chicago the Tucks permanently settled in Los Alamos, where he became Associate Division Leader of the Physics Division. While at Los Alamos, his reputation grew steadily. In the early 1950s, he led a crusade to prevent the bulldozing of Ashley Pond, currently a charming park and pond in the heart of Los Alamos. For a generation, people on the Hill termed it "Tuck's Pond." The existence of "Tuck's Mesa" to the north of the city testifies to his love of hiking with other "Wednesday Walkers." When Robert Porton, long-time head of Public Relations for LANL, needed a scientist to address visiting high school students, he invariably sought out Tuck. His enthusiasm, wit, and charm of manner never failed to reach his audience.[30]

A non-political person (his daughter later confessed that she never heard him utter a single political comment), Tuck lived totally for the world of science. In this area, however, his skills ranged far and wide: from expertise in implosion triggers to the study of ball lightning to the exploration of fusion power. He spent much of his postwar career working on Project Sherwood, a long-term experiment in coaxing commercial power from fusion. He was so well known that a letter addressed to "James Tuck, Los Alamos" found him without problem.

Brilliant and unconventional, Tuck also had an uncanny ability to arrive on the scene when things began to malfunction. Tuck stories soon assumed the status of local folklore: how Tuck went to work with mis-matched socks, Bermuda shorts, and unkempt hair; how his secretary once stapled his cuffs together because he owned no cufflinks; how he drove with the top down on his convertible in mid-winter; how he kept a barber chair in his basement where he could retreat, smoke his pipe and simply think; how he nailed a blackboard next to the dining room table so his two children could work out mathematical problems to impress visitors; how he forgot to deposit his monthly paychecks from the Lab. Los Alamites always tell these stories with a smile and a chuckle. As John Manley recalled, Tuck was "a character by just being Tuck."[31]

Over the years James Tuck's ever-present pipe, like Robert Oppenheimer's porkpie hat, assumed almost symbolic proportions on the Hill. In a certain sense, James Tuck took on the role of comic relief for Los Alamos, a position that helped moderate the tragic role that had been forced upon Robert Oppenheimer.

Almost a third of the British Mission to Los Alamos – Bretscher, Frisch, Fuchs, Peierls, Placzek and Rotblat – were émigrés or refugees from the Continent. Thus, they were "British" as much by force of circumstances as by design. Peierls had become a British subject in 1940, but Bretscher and Frisch were naturalized, called up for military service, and then exempted within a twenty-four hour period just before departure to the States. Frisch actually received his British passport only after he had boarded the ship for the United States. In a unique arrangement, Joseph Rotblat remained a citizen of Poland while he worked on the Hill.[32] These refugees, naturally, retained heavy continental accents, and a wag once remarked that the British Mission couldn't speak English.

Initially all the British scientists were to work under James Chadwick, but he soon decided to leave Los Alamos for Washington. Chadwick's decision to leave was doubtless influenced in part by family concerns. Neither his wife, Lady Chadwick, nor their twin, teenaged daughters adjusted easily to the Los Alamos environment.[33]

Lady Chadwick had never been stateside before and experienced difficulty coping with the demanding living conditions at Los Alamos. Once she invited several wives over for a formal high tea and, during the course of the conversation, delivered an attack on the primitive nature of life in the United States. This annoyed several of the American wives. As they departed, Bernice Brode whispered to her colleague that she wished Lady Chadwick could see their comfortable home in Berkeley.[34] On another occasion, Lady Chadwick expressed surprise that the British had entrusted so much responsibility for D-Day to the Americans. As Hanni Bretscher remarked, Lady Chadwick did not behave in a manner calculated to make people like her.[35]

In retrospect, the Chadwicks' move to Washington worked for the best. Lady Chadwick utilized her social skills better in the Washington political environment and the Capitol proved a much more central location for Chadwick to coordinate the details of Anglo-American cooperation on nuclear matters. He carried out these often delicate negotiations with Groves and the Pentagon and with both the Roosevelt and Truman administrations with skill and understanding. In Chadwick's delightful phrase, all had become "jam and kippers."[36] Through the endless conference they both attended, the shy, reserved Chadwick and the blunt, outspoken Groves developed a genuine admiration for each other. They became lifelong friends, and Groves visited the Chadwicks whenever he went to England after the war. Not until years later did his friends reveal that Chadwick suffered through nights of prostrating pain produced by anxieties all through his stay in Washington.[37] When Chadwick returned to England after the war, he was completely exhausted from the strain of his activities.

Chadwick's departure from Los Alamos signalled complete integration of the British scientists into the various Laboratory divisions. Egon Bretscher was secretly delighted at the Chadwicks' move, for it allowed him to leave the frustrating job to which he had been assigned to work on questions regarding fusion theory. After Chadwick's departure, Rudolf Peierls assumed his place as the head of the British Mission.

As head of the Los Alamos Mission, Peierls's new duties included sending and receiving the coded teletypes to Washington. At times P.B. Moon or James Tuck spelled him in this task. On other occasions, Peierls was called to serve as "father confessor" to members of the British team. He advised Donald Marshall after the latter's automobile accident, and did his best to console Ernest Titterton when their first daughter was born with spina bifida. Peggy Titterton had been exposed to high doses of radiation during her first trimester at the Lab. The Tittertons have always believed that this was the cause of their daughter's handicap.[38] The child's life was probably saved through the skill of the Los Alamos medical staff.

In addition to these duties, Peierls decided to keep Chadwick informed of the progress of the laboratory. Thus Peierls wrote him periodic summary reports – perhaps ten overall – and sent them on to Washington. Chadwick later termed these summaries "indispensable."[39] Strictly speaking, Peierls was violating security, for such information was not supposed to leave the Hill without special permission. In fact, it was not long before Richard

Tolman, Groves' science advisor, visited Los Alamos and sought out Peierls on the matter. Peierls expected a mild reprimand. Instead, however, Tolman told him that Groves found Chadwick better informed on Los Alamos matters than he, and wondered if Peierls would also send him copies of the Chadwick reports.[40]

The unique social and scientific environment at Los Alamos helped democratize life in a manner new to many British Mission members. Household help by maids from the nearby Tewa-speaking Indian pueblos of San Ildefonso and Santa Clara was allotted strictly according to need (usually defined as number of small children), not rank or social standing. The housing office ran on the same criteria. Hanni Bretscher recalled an incident that reflected this lack of "social ladders." Robert and Kitty Oppenheimer once joined a long queue to view a popular film at Theatre II. Instead of pulling rank and moving to the front of the line, they waited their turn just like the others.[41] Rudolf Peierls wrote that "It is an enormous pleasure . . . to be at a place . . . where work is guided by the necessity to get the best answer in the shortest possible time rather than by questions of formal organization and prestige."[42]

The Tittertons relished this social atmosphere. They astounded General Groves by praising their living accommodations. After reading about log cabins for years, they now had a chance to live in one, they told him. Groves was so delighted he gave Peggy Titterton the run of the commissary for the day.[43] Donald Marshall relished the democratic atmosphere and formed many American friendships. James and Elsie Tuck also fitted easily into this egalitarian environment. In his memoirs, Tuck remarked that he and his wife probably enjoyed the Los Alamos world more wholeheartedly than any of the other members of the British Mission.[44]

While there was complete integration in all facilities, the Americans still regarded the British Mission members as "different." While the Americans and the British spoke the same language (more or less), in the early 1940s they shared two distinct cultures. This was reflected in subtle ways at Los Alamos. The bachelors among them, such as Frisch and Fuchs, lived in the Big House and ate their meals at the lodge, instead of living in dormitories and dining at the Mess Hall, as did their American counterparts. The British often dined among themselves. Asked not to mention the phrase "Tube Alloys," the term fell into disuse, except when the Mission members talked with each other. As British citizens, they paid no US taxes on their salaries. The government asked them to avoid political discussion, and in spite of occasional exceptions, most did as instructed.[45] When Clement Attlee's Labor party defeated Churchill's wartime Coalition government in July, 1945, it came as a surprise to most Americans on the Hill. The British

Mission members had discreetly avoided mentioning the subject. Only after the election did they share their opinions on the issues involved.[46] Thus, in spite of the social integration, it was generally understood that the British were at Los Alamos as "guests."[47]

Another item that distinguished the British Mission from their American co-workers was their relative lack of funds. Mission members joked that they slept on sheets marked USED (United States Engineer District), but in truth, the wives cooked in borrowed GI pots and pans. Only Fuchs and Peierls drove their own cars. The financial arm of wartime austerity reached all the way to New Mexico. The British government paid travel expenses for the scientists but some wives and children funded part of the journey by themselves. Hanni Bretscher and family arrived in New York down to their last shilling. She did not dare be unhappy at Los Alamos, she recalled, for the family could never afford the journey home.[48]

This lack of funds was highlighted by an accident involving British scientist Donald G. Marshall. While driving without authorization, Marshall badly damaged an Army automobile. The standard procedure in such a situation was for the employee to reimburse the US government for all expenses involved. Oppenheimer believed that Marshall fell under this rule. The situation fell to Ralph Carlisle Smith's jurisdiction, however, and he decided that the circumstances of the accident dictated otherwise. He overruled Oppenheimer and declared that Marshall did not have to pay. Afterwards, Chadwick wrote Smith a letter of thanks, confessing that he foresaw no means by which Marshall could have reimbursed the government for the car.[49]

All the British Mission memoirs recall the Los Alamos years with pleasure. Kathleen Mark observed that the primitive living conditions took some getting used to, but after a while one took little cognizance of the ugliness.[50] A.P. French confessed that his time at Los Alamos was the most fascinating single period of his scientific career. French considered himself unbelievably lucky to have been a part of that society. D.J. Littler and P.B. Moon felt much the same way. Tony Skyrme recalled that most members of the British Mission "were happy, professionally and personally."[51] James and Elsie Tuck termed Los Alamos "an unforgettable and enjoyable experience" while Kathleen Mark marvelled at how well the people there got along with each other.[52] Overall, Hanni Bretscher recalled her Los Alamos sojourn as a "most marvelous time."[53] Peggy Titterton said she would not have missed it for anything.[54] As Poole later observed, mention of a Los Alamos connection opened many doors. If one were in the physics business, he recalled, being at Los Alamos was "the finest club you could ever belong to."[55]

By consensus, the social event that best personified this Anglo-American spirit of cooperation at Los Alamos came on Saturday, September 22, 1945. That night the British Mission members put on a party to celebrate "the birth of the Atomic Era." With $500 supplied by both the British Mission on the Hill and the Embassy in Washington, the Los Alamos group pooled their ration points to shop with *élan* in nearby Santa Fe. Klaus Fuchs purchased the alcohol, and people worried when he was late in arriving.

The celebration that evening reflected a distinct "British flavour." Invitations were engraved. Guests arrived in "formal" attire, many of the women in white gloves. A "footman" announced the arrival of each guest. Dinner began promptly at eight. British Mission members all had their own tables, and they invited their own guests. The Mission wives had worked for weeks at their "most secret" (British "top secret") preparation. Genia Peierls made a thick pea soup in pails, while Else Placzek and Hanni Bretscher contributed turkey, boiled ham and English potato salad. Winifred Moon's dessert of trifle became an object of considerable interest to the Americans, most of whom had never seen it before. Several hid theirs in the long table drawers to be discovered much later. Rudolf Peierls recalled carving roast beef for over 100 people.[56]

The best port wine was then liberally dispensed for a round of ceremonial toasts. The group raised glasses to the King, the President, and, especially, to the health of the Grand Alliance. The hall was absolutely packed.

After the meal came entertainment. All members of the British Mission on the Hill (the Chadwicks did not come) helped stage a "British style pantomime" where a narrator told a story that was acted out by silent performers. Titterton accompanied the action on the piano to the tune of "Atcheson, Topeka and Santa Fe." Entitled "Babes in the Woods," the play told the saga of the British Mission in Los Alamos and how "Good Uncle Franklin's forces" outwitted "Bad Uncles Adolph and Benito." The dialogue contained a number of *double entendres* about the project that probably eluded several of the American wives.

Through several short sketches, the members of the Mission lampooned life in wartime Los Alamos. Censorship was parodied when an actor dropped a letter into a slot – only to have it fly out back at him. Otto Frisch, dressed as an Indian maid, solved the problem of house cleaning by stuffing everything in a closet, breaking crockery on the floor, and then pouring himself a stiff drink. James Tuck, dressed in a red devil's outfit with a bulb on his tail that blinked on and off, poked fun at the pervasive security arrangements. He forced Philip Moon, who wore his footprint as an identification badge, to chew up and swallow all documents that he removed from the safe. He then sentenced Moon to a thousand years in jail for security violations.

The climax skit came when the Mission re-enacted the Trinity Site explosion. A stepladder represented the steel tower and a bucket on top was "detonated" amidst flashes of light with appropriate sound effects. At the climax moment, George Placzek flicked his cigarette into the bucket. The "explosion" was not entirely comprehensible to many of the observers. Then came an evening of dancing. Peggy Titterton confessed to eating and drinking too much that night. In addition, one junior British scientist began throwing light bulbs from the balcony – one of which just missed Kitty Oppenheimer – and Peierls had to suggest that he go home to bed. Bernice Brode termed the party one of the finer moments in the history of the Grand Alliance. Ralph Carlisle Smith recalled it as "one of the best parties ever held at Los Alamos."[57] As E.W. Titterton noted, the farewell party soon achieved the status of Los Alamos legend.[58]

After the party, the British Mission members began their preparations for departure. They wrote up their final reports, and turned in all equipment and documents. A few weeks earlier, the Mission members had put together a several-page policy statement on the implications of the new atomic world. After it was signed by everyone, they sent this document to Washington, to be forwarded to Whitehall. "Only people who had experienced Alamogordo could realize the enormity of the meaning of that flash of light," Titterton recalled. "We felt it our duty to bring it to the notice of the people who would have to live with this." Titterton believed the document helped influence the "deep thinkers" in London.[59]

Groves, who by now had emerged as a bit of an Anglophobe, encouraged all foreign personnel to depart Los Alamos as soon as possible.[60] From Washington, Chadwick advised Moon that all British workers should be certain to take with them all their notebooks and copies of all technical reports and memoranda with which they had been concerned. "We cannot leave all our information behind and go home with empty hands," Chadwick warned. Future nuclear work in England depended on this, and the subject was foremost in many of his letters.[61] Indeed, the British left with extensive knowledge of all the work, except, perhaps, the Oak Ridge reactors and the details of plutonium production at Hanford. Sir John Cockcroft later noted that "we acquire[d] an almost complete knowledge of [the bomb's] technology."[62] Historian Margaret Gowing observed that, among all of them, they knew "most of what there was to know" about weapons manufacture.[63] Norris Bradbury admitted that in 1946 the British knew everything the Americans did.[64] Los Alamos had demonstrated a rare example of complete scientific and social integration. The British and

American scientists on the Hill demonstrated little of that "remarkably close and yet particularly strained" relationship that had characterized so many other aspects of the Anglo-American wartime alliance.[65]

4 The Aftermath

Immediately after the war, the entire Manhattan Project fell into limbo. It remained in a generally confused state until the passage of the Atomic Energy Act in 1946 revitalized it with the creation of a new structure, the Atomic Energy Commission (AEC). The AEC officially took over on January 1, 1947. During this eighteen-month period, General Groves presided as best as he could over a sprawling, disintegrating organization. Much to his dismay, he watched the atomic scientists mobilize to defeat the proposed May-Johnson legislation (regarding the AEC), which they criticized as giving too much power to the military. Instead, the scientists threw their support behind a law drafted by Connecticut Senator Brien McMahon, the McMahon Act, which placed the chief power in civilian hands. Congress debated the issue for almost nine months. After numerous compromises, one of which established a powerful Military Liaison Committee, Truman signed the McMahon Act into law on August 1, 1946.[1]

The discussion on this legislation proved intense. One observer declared that Truman's support for civilian over military control of atomic power would rank as one of his greatest achievements. Reporter Raymond Gram Swing said that Truman's recommendations about the regulation of the atom were more important than any of Franklin Roosevelt's decisions. Secretary of the Navy James Forrestal called Truman a "rocklike example" of civilian control over the military. Historian Necah Furman has argued along similar lines. She termed the legislation to control atomic energy "in some respects as revolutionary as the scientific discovery that caused its creation."[2]

While this battle was raging in the press, Chadwick oversaw the final departure of the Los Alamos Mission. Representatives from the British government arrived in the nation's capital to recruit personnel for the proposed British Atomic establishment at Harwell. Several of the junior Los Alamos people took the train to Washington to be interviewed for possible jobs in the British atomic program. Before departing, Peierls also gave a series of lectures on hydrodynamics for the short-lived "Los Alamos University." From December, 1945, onward, the British Mission departed the Hill in a steady fashion. M.J. Poole left as early as he could to marry his fiancée, who had remained in Britain. Donald G. Marshall also left early, largely because of general unhappiness with Los Alamos. Marshall's expertise lay in explosives research, not nuclear physics, and he spent numerous hours wading through mud in remote canyons where such work was done. The Peierls also departed in January. At Norris Bradbury's urging, Fuchs remained until June 15. Most of the Mission hoped to move back into the academic world as soon as they could. As A.P. French recalled, "Once the principles of the bomb were well understood, then really the physicists' contribution was over. It was engineering."[3]

But some British scientists – Mark, Bretscher, Penney, Titterton, and Tuck, among them – remained on the Hill several months longer. In the spring of 1946, those still present dined at neighboring "celebrity" Edith Warner's house, "a most unique experience," as one observed.[4] The British scientists watched the American political battles over postwar atomic control with interest. They also applauded the November 15 joint declaration by President Harry Truman, Canadian Prime Minister Mackenzie King, and British Prime Minister Clement Attlee on their plans to prevent future wars and share scientific information.[5]

Until the AEC could be created, however, the American Military remained in control of the nuclear enterprise. Late in 1945, the military decided to stage two further atomic tests on the Pacific island of Bikini, code-named "Operation Crossroads." Thus on July 1 and July 15, 1946, the US detonated atomic bombs "Able" and "Baker." A proposed third test, "Charlie," was cancelled because of rising concern over radiation danger. "Able" was dropped from the air while "Baker" was detonated under the waters of the Bikini atoll lagoon. The ostensible purpose was to discover how effective atomic bombs would be against ships at sea. The real purpose may have lain elsewhere, either in inter-service rivalry or in a simple raw display of power.[6]

"Operation Crossroads" turned into the major media event for 1946. Harry Truman actually postponed the test for six weeks so that several Congressmen could fit it into their schedules. Reporters from all over the world descended on Bikini, and a special ship was set aside just for them. The journalists asked so many questions that they thoroughly annoyed the scientists. Titterton became especially exasperated. The "Able" explosion at Bikini was probably the most photographed event of the year. Overall, the Crossroads test served less to advance bomb technology than to demonstrate its results to the world. The island of Bikini is still a household word, but only because a swimsuit manufacturer borrowed the name for his product.

The Crossroads test also raised some delicate questions regarding Anglo-American cooperation. Many American scientists were unenthusiastic about the Crossroads project, so the British were asked to help in the test preparations. In all, eight British scientists participated in these tests, three from Los Alamos and five from the British Isles. Penney's skill in blast measurements, Tuck's in radiography, and Titterton's with electronic timing were all essential to Crossroads. Titterton actually delivered the final countdown over the loudspeaker.[7] Groves requested that Los Alamos not call attention to the presence of British personnel in the Crossroads series. Consequently, when a group of newsmen interviewed Titterton, they promised not to mention that he was British. Instead, they referred to him simply as "the voice of Abraham." In spite of the extensive press coverage, the reporters did not stress the British participation. Even the official report of the test, *Bombs at Bikini*, fails to mention the British contribution.[8]

Groves made this rather strange request because of the provisions of the upcoming McMahon bill. Although the new regulatory agency would not officially take over supervision of atomic matters until January 1, 1947, the August legislation would drastically affect all British–American scientific exchange. Section 10 of the bill made the distinction between "basic scientific data" and "technical processes." The former could be shared with other nations but not the latter. After passage of the Act, only US citizens could have access to "restricted data" (itself a new classification). "Restricted data" was defined as all data concerning the use of atomic weapons, production of fissionable material, or the use of this material to produce power. In fact, the Atomic Energy Commission reviewed the Top Secret British Report on Crossroads to insure that it did not violate any of these provisions.[9]

Immediately after the war, Truman had warned the British that America was not about to reveal "engineering and production know how" on atomic matters "any more than we make freely available our trade secrets."[10] On August 1, this position became the law of the land.

Passage of the McMahon Act raised severe problems for the British scientists still at Los Alamos. What scientific documents could they continue to examine? At Los Alamos, Ralph Carlisle Smith argued that the British scientists on the Hill should be allowed access to all documents and reports that they had utilized prior to August first. Norris Bradbury, recently appointed head of the Laboratory, suggested that Smith ask Groves for his opinion on the matter. Groves replied with a far more strict interpretation: no foreign scientist could have *any* access to restricted data, regardless of his previous status. It was the law.[11]

This new situation proved both frustrating and embarrassing to all concerned. In February, 1947, all British Mission employees had to leave the Technical Area. Carson Mark wrote several letters to express his dismay at the new restrictions, but to no avail.[12] Others, such as Egon Bretscher, returned to England in disgust. "The whole thing was done as awkwardly as possible," recalled Carson Mark. "A person was denied access to his own reports."[13] Ernest Titterton remained in Los Alamos until the spring of 1947. He could continue to work as a foreign national because his specialty was electronics and most of the electronic data remained unrestricted. This was a subterfuge, however, for the detonation of plutonium weapons demanded microsecond electronic precision. Eventually, however, Titterton also had to leave. When he finally departed the Hill on April 12, 1947, the British Mission to Los Alamos officially closed its books.[14]

The passage of the McMahon Act creating the Atomic Energy Commission inaugurated a series of major problems for US–UK atomic relations. The British had assumed that the (almost) complete nuclear cooperation would continue in the postwar era. The United States, however, declined to maintain their "special relationship" with England while attempts were underway in the United Nations to establish a viable system of international control over nuclear weapons. American officials feared that a close atomic relationship with Britain might endanger a genuine multinational plan of regulation.

Written largely by Oppenheimer, the American scheme for internationalization of nuclear weapons was termed the "Baruch plan" after the distinguished Ambassador, Bernard Baruch, who introduced it into the United Nations. The Soviets mounted objection after objection to the Baruch plan, however, and after over fourteen months and almost 200 conferences, the idea died. Even after this failure, the US still showed little willingness to share atomic knowledge with its former British allies. Only in the area of joint gathering of uranium supplies was there real cooperation.[15]

The shift from informal wartime cooperation between Roosevelt and Churchill to formal diplomatic agreement by the Congress and Parliament proved painful. The two semi-autocratic heads of state had thoroughly enjoyed their top-level discussions. Each remained fascinated by his counterpart, for they understood each other as only two patricians could. (Churchill's mother, of course, was American.) The real "special relationship" between the two nations ultimately lay in this wartime Roosevelt–Churchill understanding.

Harry Truman, however, was cut from a different piece of cloth. His small-town Missouri political training gave him a very different view of the world. The comments in his Potsdam diary were often very critical of Churchill. When Churchill visited Fulton College in Missouri for his famous "iron curtain" speech in 1946, relations thawed somewhat, and at the poker table they became "Harry" and "Winston." But Churchill was out of office then, and perhaps more sanguine about US–UK relations. Even when he became Prime Minister again in 1951, the renewed Churchill–Truman understanding never approached the depth of the Churchill-Roosevelt alliance.[16]

On September 21, 1945, Truman asked his cabinet for their views concerning control of the atomic bombs, especially in reference to disclosing data to the Soviet Union. The responses varied, but the majority felt that the United States should keep "the formula" to itself. Secretary of Agriculture Clinton Anderson doubted that the Soviets possessed the "know-how" to make such a weapon by themselves. Tennessee Senator Kenneth McKellar agreed, but with an economic emphasis. "We do not ourselves know the value or worth of this formula at this time," he warned. Truman's Secretary of State James Byrnes felt the same way. Although the scientists argued for internationalization, Byrnes remarked that, on this issue, they "were no better informed than he was on the construction of the bomb."[17] Anderson reminded the President that he had talked to over 200 ordinary citizens and none of them advocated sharing the "secrets." An October State Department report found that most American politicians also opposed sharing atomic information and control.[18] Unsolicited letters to the President confirmed this view.

The responsibilities of controlling the new atomic power lay heavily on Truman's shoulders. In August, 1945, he termed America "the most powerful nation in the world – the most powerful nation perhaps in all history." Two months later he called the US monopoly of atomic weapons "a sacred trust."[19] Although he listened to the scientists' predictions about the value of internationalization, he was very comfortable with the status quo.

Historian Francis Duncan has shown that the Congressmen who drew up the McMahon Act had been almost totally ignorant of the role that the British had played in the Tube Alloys/Manhattan Project story.[20] They remained unaware of the September 18, 1944 Hyde Park *aide-mémoire*, for example, that called for full cooperation in both the peaceful uses of atomic energy and the military application after the war. (The document had been misfiled.) Until the American original surfaced, Groves actually considered it a British forgery. On May 5, 1947, at the initial meeting with the Joint Congressional Committee on Atomic Energy, several senators confessed they were unaware that British scientists knew how to make a bomb. The Foreign Affairs Committee members did not know that either.[21] Even Senator McMahon remarked that if the committee "had seen this [Quebec] agreement, there would have been no McMahon Act."[22]

Meanwhile, a significant number of scientists, including Hans Bethe, Marcus Oliphant, and others, predicted that any industrialized nation (specifically, the Soviet Union) would have the bomb within three to five years. Sir James Chadwick expressed this view on the front pages of the

New York Times. "I think this is a very serious point," he told the nation on August 13, 1945.[23]

Truman, however, was not convinced. While the President did not exactly distrust the atomic scientists, he had had no previous dealings with them either. Moreover, he considered scientists as simply one voice among many in a democracy. "We need men with great intellects, need their ideas," Truman once told AEC Head David Lilienthal. "But we need to balance them with other kinds of people, too." A former soldier himself, Truman listened to General Groves. He felt far more comfortable with the military point of view that opposed "sharing" atomic secrets (especially if public opinion seemed to agree). He also expressed considerable scorn for the Soviets. He argued that the technical "know-how," the cost and the materials of an atomic weapon were beyond their reach. In 1946, he gloated that Russia "could not turn a wheel in the next ten years without our aid."[24] Thus, keeping the atomic monopoly became an integral part of Truman's administration. As recent studies have shown, the armed services, especially the Air Force, based their early postwar military strategy almost exclusively along the lines of an American atomic monopoly.[25]

The British were understandably dismayed by the McMahon Act's exclusion clauses. Even worse, they felt betrayed. On June 7, 1946, Prime Minister Clement Attlee wrote Truman a five-page letter detailing the early history of the atomic age and the British contribution to it. He reminded Truman of the Hyde Park agreement between Churchill and Roosevelt. Former Secretary of State Henry Stimson had also advised Truman that Britain "has the status of a partner with us in the development of this weapon."[26] But by 1946 Stimson had been replaced by Byrnes, and Attlee's pleas also fell on deaf ears.

Moreover, the British emphasis on how much they had helped the Americans in the past had minimal impact on American negotiators. The Americans seemed far more interested in the British potential for *future* cooperation.[27] Thus, from 1946 forward, the Americans declined to share their nuclear data with the British. Historians Hewlett and Duncan and Margaret Gowing have each emphasized the awkwardness of this. American behavior, noted John Cockcroft, was full of "evasiveness, indecision, tergiversation and downright mendacity.[28]

The problems ranged from the basic to the petty. In June 1948, Cyril S. Smith, a prominent expert in the metallurgy of plutonium, was prohibited from even discussing the subject when he met with a group of British scientists at a conference. In another incident, American physicist Luis Alvarez heard a complaint from William Penney that his men had not yet been able to discover what crucible material Los Alamos had used to melt plutonium in. When melted, the plutonium dissolved or interacted with all the crucible ceramics the British had tried. Both Alvarez and Penney knew they could have found the data in five minutes in the library at Los Alamos. Thus, Penney had to waste valuable metallurgical man hours in the search. Alvarez knew the answer, but he could not tell Penney.[29]

All this became both embarrassing and frustrating. Basic scientific research data, the US would provide, but it would share no technical data with the British or with anyone else. As Assistant Secretary of State Dean Acheson privately told a member of the British embassy, "Although we made the agreement [to share nuclear knowledge] we simply would not carry it out . . . things like that happen in the Government of the United States due to the loose ways things are handled."[30]

Unbeknownst to the British people – and probably to the American government as well – the British had already made plans for their own atomic enterprise. In fact, their plans had been laid well before passage of the McMahon Act.[31]

At the conclusion of the war, the future American relationship with Europe was by no means clear. The Marshall Plan and NATO both lay several years in the future. Truman's abrupt termination of Lend Lease, in August 1945, plus the relatively modest 3.75 billion dollar British aid package, were not encouraging signs.[32] Attlee's government feared that America might withdraw into isolation once again. All these items contributed to the feeling that Britain must develop the atomic bomb for herself as soon as feasible.

When the members of the British Mission to Los Alamos returned to Europe and England, they assumed a variety of professional positions. Bohr resumed the leadership of his Copenhagen Institute. Peierls and Bretscher accepted academic posts, the former at Birmingham and the latter at Cambridge. Moon also returned to Birmingham. After a brief sojourn at Harwell, Otto Frisch was appointed to a chair at Cambridge. Several of the younger men, such as French and Poole, returned to graduate school to finish their doctorates. Only William Penney entered full time into British weapons work.

Yet the psychological dimension of the Los Alamos experience should never be discounted. Among them, the British team knew all the technical details of weapons manufacture. Although the McMahon restrictions proved constantly annoying – British negotiators would try for over a decade to revise them – they were never viewed as insurmountable. Engineer Christopher Hinton, who oversaw the industrial dimension of the British bomb project ("the Brunel of the twentieth century") actually declared the McMahon Act a blessing in disguise. He argued that exclusion from

American data forced the British scientists to innovate in several areas. Chadwick felt the same way. "Are we so helpless," he said, "that we can do nothing without the United States?"[33]

Thus, shrouded in secrecy (yet another legacy of the Tube Alloys/ Manhattan Project) the British scientists worked for seven years on their own nuclear weapons program. Harwell became the Chief Research Center, Windscale on the Cumbria Coast served as the plutonium factory, and Aldermaston in Berkshire as the weapons laboratory. Unlike the States, where the military supplied most of the finances, British funding came from several ministries, largely civilian organizations. Initially, the chief British problem was not money but permission to build buildings, for steel and other building materials were all in very short supply. The British felt they had to re-establish their position in basic and applied science. The development of an independent nuclear deterrent provided that opportunity. In Margaret Gowing's phrase, this became "one of the most successfully executed programmes in British scientific and technological history."[34]

Although Penney and his staff were considerably piqued that the Soviet Union became the world's second nuclear power in August of 1949, they brought Britain into the nuclear club in October of 1952. Five years later they successfully detonated a hydrogen bomb on Christmas Island in the South Pacific. (All this will be discussed in the next chapter.)

The creation of a British independent nuclear deterrent, plus the hardening of Cold War ideologies, made the Americans rethink the McMahon Act exclusion clauses. In 1954/5 Congress amended the Atomic Energy Act. The new provisions allowed the British to receive some restricted data on atomic weapons. Although the revisions were quite modest, they opened the door for the major breakthrough that came three years later.

The final catalyst for this change came on October 2, 1957, when the Soviets successfully launched the world's first satellite in *Sputnik*. ("Something equivalent to Pearl Harbour," Prime Minister Harold Macmillan observed.) *Sputnik* made the Americans doubt that they were "winning" the arms race.[35]

Immediately afterwards, President Dwight Eisenhower, whose friendship with Macmillan dated back to the war years, urged the Prime Minister to fly to Washington to discuss the McMahon Act. Secretary of State John Foster Dulles had termed the act "obsolete" and Eisenhower told Macmillan he was ashamed of it as "one of the most deplorable incidents in American history."[36] The timing of Macmillan's visit proved propitious, for both Congress and the American public had recently been charmed by the arrival of Queen Elizabeth II to participate in the 350th Anniversary of the founding of Virginia. Extensive press coverage of her visit, remarked the British Ambassador Harold Caccia, had "buried George III for good and all."[37]

Macmillan basked in his good fortune, astounded at how swiftly the American government could move if it chose to. In July, 1958, Congress replaced the McMahon Act with the Atomic Bilateral Agreement. The period of exclusion had ended. This was, in the Prime Minister's words, "the great prize."[38]

Historians Lorna Arnold and John Simpson have suggested that the Americans received more than their fair share of the 1958 agreement. After 1958, they had, most of all, access to William Penney and his weapons laboratory. During the initial meeting of the US–UK panel of expert negotiations, Edward Teller remarked that the twelve-year separation had allowed the British scientists to reach the same level of physics understanding that the Americans had achieved. While the greater American industrial enterprise had produced more sophisticated engineering, the British had also discovered certain technical innovations that the Americans were pleased to incorporate into their own program.[39]

Simultaneously, however, on October 10, 1957, a major fire broke out at Britain's nuclear reactor at Windscale, Cumbria. Radioactivity released into the atmosphere spread across England, Wales and Northern Europe. Milk was banned for a 200 square mile area around the plant. Eventually, the damaged piles were covered in concrete, never to be entered again. Sir William Penney immediately prepared a thorough report on the incident in which he condemned both the faulty instrumentation and the actions of the Windscale staff. While Macmillan read the report carefully, he also did his best to play it down. The Prime Minister feared that the accident might cast doubts about British nuclear competence, and undermine the negotiations.[40] As this was the world's first nuclear accident, it passed without creating major public furore. It had no effect on the McMahon Act revisions, and things proceeded as planned in that regard.

With the 1958 agreement, Britain became the *only* nation with which the United States would exchange both technical information as well as fissile material. Although there were frequent quarrels over potential *commercial* use of data exchanged, these were eventually ironed out in endless meetings by experts on both sides. Of course, commercial dimensions always took a back seat to military ones.

The military coupling interwove the two defense systems in a number of areas. The two sides began to share technical data from all nuclear tests. The first joint test occurred in Nevada in 1962 and when the Americans utilized Christmas Island that same year, the British viewed all test results. Current British warheads, however, are of indigenous design, not simply copies of their American counterparts.[41] But experts from both sides often meet to exchange views on the matter. After 1958 then, Great Britain clearly harnessed herself to the American nuclear juggernaut. The results of this coupling have yet to be decided.

As historians have recently noted, the ensuing US–UK alliance was played out on the stage of shifting postwar relations. Britain was a power in decline in every realm except, perhaps, the moral.[42] She had lost one-quarter of her national wealth and her empire in Asia as well. America, on the other hand, had pulled herself out of a depression and had begun an uncomfortable new role as a superpower. Suddenly America had new responsibilities in the Pacific, East Asia, and the Middle East, where she had previously been subordinate to England.[43] At present, all American defense plans contain a British component. These issues continue to raise political hackles in the British Isles, among those who dislike being cast as "the fifty-first state." A "decoupling" of the special relationship, however, would have incredible consequences for both nations. It might not even be possible. Although there have been several modifications of the 1958 agreement since, the present-day Anglo-American military nuclear integration still remains intact. This, surely, is the major legacy of the British Mission to Los Alamos.

5 Varieties of the British Mission Experience

The story of the British Mission to Los Alamos involves more than simply a tale of Anglo-American cooperation and scientific breakthroughs. It is also a story of scientists who returned to England to forge the main outlines of the postwar nuclear world. Although they shared a similar Los Alamos experience, they held far from identical views on the major issues of the day. In fact, the British scientists eventually fanned across the entire political spectrum: from the bulwarks of the Anglo-Australian nuclear establishment to Soviet agent, with the majority falling somewhere in between.

The saga of Soviet spy Klaus Fuchs will be discussed in Chapter 6. This chapter will focus on the careers of four of the most distinguished British Mission residents on the Hill: Joseph Rotblat, William Penney, Ernest W. Titterton, and Niels Bohr. Collectively, these men articulated the great concerns of the world in the immediate postwar era.

From the beginning, experimental physicist Joseph Rotblat held a unique position in the British Mission to Los Alamos. Born November 4, 1908, in Poland, Rotblat was working at the Radiological laboratory in Warsaw when he first heard of the fissioning of uranium. In April of 1939, the Polish government arranged for him to study for a year with James Chadwick in Liverpool. But the war broke out in September and Rotblat could not return home. He was stuck in England for the duration.

Initially, Rotblat expressed doubts about the morality of working on a project that might lead to the production of weapons of mass destruction. The fall of Poland ended his qualms, however, and he threw himself wholeheartedly into the Liverpool experiments. Rotblat has always considered himself as one of the first people to evolve the concept of "nuclear deterrence." He worked on the bomb, he said later, because he held "the belief that if Germany made the bomb, the only way to prevent its use against us would be if we, too, had it and threatened to retaliate."[1]

Over time, however, he began to doubt even his own theory. In 1981, Rotblat concluded that he had been in error. If Hitler had developed the bomb, he probably would have utilized it during the final days of the Third Reich regardless of any potential retaliation. Drawing on German mythology, Hitler would have rejoiced in departing the world via a *Götterdämmerung*.[2] Nuclear deterrence worked only when both sides operated on the same rational premises. The plan would be ineffective against a madman.

In early February of 1944, Rotblat was selected as one of the team to move to the States. At Chadwick's urging Rotblat was designated a "technical scientific officer" with the British Supply Council. On February 28, he was transferred to the British Supply Council of North America in Santa Fe.[3] Since he hoped to return to Poland after the war, he retained his Polish citizenship throughout his time in Los Alamos. This proved a unique arrangement. It also troubled American security.

Rotblat's first impression of Los Alamos was that the Hill was the ideal place for science. Coming from Warsaw via Liverpool, where both materials and apparatus were in short supply, he was delighted to discover that he could get anything he wanted for his work with no difficulty. He also relished the social environment, in which the world's best physicists discussed their discipline, both on and off the job. The sunshine and mountains also proved a welcome respite from Liverpool's cold, fog, and rain.[4]

After his arrival, Rotblat roomed briefly with the Chadwicks on bathtub row. Then, after an initial period of indecision, he was added to Robert Bacher's team, where his specific assignment was to establish how many gamma rays were emitted when a neutron was captured by a U^{235} nucleus. Later Rotblat was assigned to R Division, under Robert Wilson, where he did experiments with the cyclotron.

He found, however, that even the heady atmosphere of Los Alamos could never quite overcome two areas of nagging doubt. First, throughout most of 1944, when the war was still going badly for the Allies, his mind continually raced back to Poland and the uncertain fate of his wife, Modesa, and their small son. He also had a brother who was fighting in the Russian Army. All through his stay on the Hill, Rotblat worried constantly about the safety of his family.

As Rotblat recalled his feelings in a later interview, he also began to have increasing doubts about the Project. In March 1944, he understood General Groves to say: "You realize of course that the real purpose of making the bomb is to subdue our chief enemy, the Russians."[5] He termed the shock of this conversation as "incredible."[6] Thus, in the late fall of 1944, Joseph Rotblat requested permission to leave Los Alamos and the Manhattan Project.

Rotblat's decision to depart Los Alamos doubtlessly stemmed from mixed motives. From the mid-1980s forward, however, he has emphasized his growing discomfort with the moral dimension of the Project as his main reason for quitting.

If we may accept this, Rotblat's decision to leave put him in a minority of scientists who voiced early doubts about the morality of nuclear weapons. From the beginning, physicist Lise Meitner held this opinion. She remained in Sweden for the duration and always refused to work on what she termed "weapons of destruction." German refugee physicist Max Born also remained in Edinburgh, largely because his Quaker wife had persuaded him not to do war work.[7] After Hiroshima and Nagasaki, James Chadwick confessed that several of his British colleagues had refused from the start to participate in the bomb project. They feared creating "a planet-destroying monster."[8]

Changing one's mind in the middle, however, was unique. Once people entered the gates of Los Alamos, everyone assumed they would stay until the end of hostilities. Military men who were transferred out of Los Alamos were usually sent to Alaska, New Caledonia, or some other isolated region. For a top-level scientist to leave the technical progam was unusual. Felix Bloch and Joseph Rotblat were the only scientists of established reputation to do so. After a brief sojourn at Los Alamos, Felix Bloch moved to the Radio Research Laboratory at Harvard to work on radar problems. "I quit [Los Alamos] for many reasons," Bloch said later in a rare interview, "but one of the reasons was that I didn't think it [the bomb project] was going to decide the war against Germany."[9] (Rumors abounded, at the time, however, that Bloch left primarily because Robert Oppenheimer had picked Hans Bethe to lead the Theoretical Division.[10])

Rotblat remained very close to Chadwick, and he often shared his concerns with him. After many conversations, Chadwick suggested a plan. He recommended that Rotblat let it be known that the real reason for his desire to leave Los Alamos was his concern for his wife and family in Poland. Stories to this effect soon circulated on the Hill. Thus, in early December 1944, Rotblat became the only British Mission member permanently to leave Los Alamos before the war ended. He sailed for Liverpool at Christmas 1944, under strict instructions not to speak with anyone about his work.

Rotblat's 1944 departure from the Hill inaugurated one of the strangest rumors to emerge from the British Mission experience. It received considerable circulation in Los Alamos. According to the recently released FBI records, the source for this tale was Mrs Mary Eileen O'Bryan of Santa Fe. O'Bryan was a middle-aged woman who had been friendly with Elsbeth Grant, an attractive heiress in her mid-twenties who had moved to Santa Fe in hopes of restoring her failing hearing. By chance, she had earlier met Joseph Rotblat in Liverpool, and they became reaquainted in New Mexico.

Against all regulations (although he told Peierls), Rotblat visited Grant every other Sunday in Santa Fe.

Mrs O'Bryan later claimed that one time Elsbeth Grant had consumed too much alcohol and confessed to her a strange tale: Rotblat disliked his work at Los Alamos and asked instead to be transferred to Berkeley. This was refused, because it would give him too much knowledge of the Manhattan Project. So, Rotblat planned to return to England, enlist in the Royal Air Force, and then parachute out over Poland or Russia. Once there, he would seek out the proper governmental authorities and give the Russians all his knowledge of the British–American atomic energy program. In so doing, he would save the world from another war.[11] Rotblat expressed fear that America would try to wage war on Russia if she alone had possession of the atomic bomb. Mrs O'Bryan said that Grant had told her that Rotblat was in complete sympathy with the Russians, that he had no use for religion, and that he had marvelous ideas about the brotherhood of man.[12] Grant agreed to help him in this enterprise, chiefly by founding a Communist Party cell in Santa Fe. Rotblat left her a blank check to aid the establishment of this operation.

Several forms of this bizarre tale circulated on the Hill. In 1950, FBI agents tracked down Elsbeth Grant, then living in California, and interviewed her extensively. Now married to a "would-be artist," William Howard Bopst, and a mother of three, Grant denied the story entirely. Rotblat left Los Alamos because (a) he felt shut in, (b) there were few foreign nationals there, and he felt watched every moment, (c) he didn't believe in the goals of the project, and (d) he wanted to play a more active role in the war. He had, indeed, left her a blank check, but it was for her to send him goods not available in wartime England. As she could not supply the goods, she eventually tore the check up.[13] (Another British Mission member, however, remembered that Rotblat's chief complaint at Los Alamos was that he had nothing to do.[14])

When Rotblat left Washington for New York to join his Liverpool-bound ship, Chadwick helped him load his trunk onto the train. After he arrived in New York City, Rotblat looked for the trunk, but it had disappeared. He remains convinced to this day that Army intelligence confiscated it. For the remainder of the conflict, Rotblat attempted to establish an electronics research group in Liverpool. Only after Hiroshima and Nagasaki did he speak out about his feelings.

For almost forty years, Rotblat declined to discuss publicly his reasons for departing Los Alamos. He first wrote of the experience in 1985, for the 40th anniversary issue of *The Bulletin of the Atomic Scientists*. There he stated that he decided to leave Los Alamos when it became clear to him that

Germany had not succeeded in developing atomic weapons of her own. Rotblat also criticized his colleagues who remained with the project, suggesting that they feared any opposition to nuclear weapons might harm their later careers.[15] His biographical sketch in *The International Who's Who* does not mention his Los Alamos experience.

After the end of the conflict, Rotblat became deeply concerned over the postwar international situation. For several months, he publicly advocated a complete moratorium on all nuclear research for three years. He gave talks to this effect in Cambridge, Oxford, London, Manchester, and elsewhere. He met with vigorous opposition on this, however, mostly from left-wing groups. They viewed a moratorium on nuclear research as an attempt to exclude the Soviets from atomic weapons. Eventually Rotblat dropped the idea, and turned, instead, to help establish the Atomic Scientists Association, the British counterpart to the American Federation of Atomic Scientists. The goal of both groups was to inform the public of the causes, consequences, and risks of the new nuclear discoveries.

To aid this educational effort, Rotblat devised a number of working models of various breakthroughs in atomic history. He arranged for this exhibit to be placed on two railway carriages, later dubbed the "Atom Train." The "Atom Train," really a science museum on wheels, toured all through the British Isles, and even made a trip to the Continent. Wherever it went, it was well received.

After his Los Alamos experience, Rotblat changed careers. He moved into the field of nuclear medicine and devoted the rest of his working life to the medical aspects of physics in general and nuclear physics in particular. He held a chair at the University of London and was attached to the Medical College of St Bartholomew's Hospital in Charterhouse Square. In 1983, the London *Times* acknowledged him as "a world authority on radiation."[16]

Joseph Rotblat garnered an American following in the spring of 1955 when he successfully explained the reasons for the widespread fallout from the States' March 1, 1954 BRAVO H-bomb test at Bikini atoll. In the *Bulletin of the Atomic Scientists* (May, 1955) he explained the fission–fusion–fission character of BRAVO, suggesting that the U^{238} "wrapper" was responsible for the majority of the fall-out. He concluded his essay by observing that "There is something particularly sinister about a bomb which is so designed as to poison the whole world with radioactivity."[17]

As the 1950s arms race and stalemate on nuclear matters continued, Rotblat also assumed another public role: British atomic gadfly. An inveterate writer of letters to the London *Times*, Rotblat regularly reminded its conservative readers that the scientists had a duty to cooperate with the press in keeping the public informed on nuclear matters. For over four decades he scoffed, pronounced, chided, and warned the British public.[18] During the 1950s he wrote articles and gave speeches to caution against the risk of genetic damage if above-ground nuclear testing were allowed to continue. He denounced many a governmental report on nuclear matters, always insisting on absolute scientific objectivity.[19]

Over the years, Joseph Rotblat also assumed a high profile as a commentator on the question of nuclear power for Great Britain. The power issue remained, perhaps, one of the most perplexing for the energy-short island nation to resolve. As writer John Walker observed in 1977, "Only the economists' arguments about the causes of inflation or those of the politicians about [Britain's] joining Europe could have produced more confusion in the minds of the general public than the arguments about nuclear power."[20]

Initially, Rotblat favored nuclear power for Britain. Nuclear generators involved risks, but so too did every other power source. He argued that British society had to reach a decision about the level of risk it was willing to accept.[21] But by the early 1980s, he began to shift positions. He now argued for a nuclear-free world: complete nuclear disarmament and the total elimination of all nuclear reactors. Utopian though this might sound, he claimed that world security depended on it.[22] It has become increasingly clear, he wrote in 1981, that "the peaceful and the military atoms are one and the same thing. Plutonium from energy reactors is now being diverted for weapons and more countries seek to procure nuclear energy facilities as a way to the acquisition of nuclear weapons." Thus, he opposed all fast breeder reactors, because of the dangers of proliferation, and urged stricter safety standards for all existing reactors. "A peaceful world – if one is ever to come into being –" he said, "will have to meet its energy needs from sources other than nuclear power."[23] Rotblat's opponents countered by attacking his "well-known anti-nuclear prejudices," and dismissed him as an "anti-nuke."[24]

When it was suggested that Rotblat had become the "conscience" of the British Mission, he dismissed the idea as "a bit grandiose." He did admit, however, that he was the only one who departed Los Alamos when the original intention of the program – to beat the Germans to the secret of the atomic bomb – proved no longer valid. Yet, he also observed, "Not being an absolute pacifist, I cannot guarantee that I would not behave in the same way, should a similar situation arise."[25]

Although surely mixed with other motives, Rotblat's 1944 departure from Los Alamos remains symbolic. He became one of the first scientists to raise the question of the morality of work on weapons of mass destruction. Because of the secrecy of the Manhattan Project, Rotblat could not take the issue into the public realm. All he could do was respond as an individual. In the 1970s and 1980s, this question has received considerable publicity, both over the issue of continued weapons research and the proposed Strategic Defense Initiative. The key voices have often been from religious groups – the Roman Catholic Church, the historic peace churches (such as the Quakers and Mennonites), the Seventh-day Adventists, and Jehovah's Witnesses. This issue is not easily resolved; the question has to be answered on an individual level. Yet it was first articulated when Joseph Rotblat left the British Mission on the Hill for Liverpool in December of 1944.

The career of William George Penney proved quite different from that of Joseph Rotblat's. Son of a sergeant-major in the Royal Army Ordnance Corps, Penney financed his brilliant academic career through scholarships. When the war broke out, he was a very young professor at Imperial College in London. He remained in London to coordinate the British and American research on the Allied bombing of Germany. Through this effort, he became the world's expert on bombing and blast waves. He was sent to Los Alamos because of his mastery of this field.

As soon as Penney arrived in Los Alamos, in the spring of 1944, he was asked to report to the director's office. Oppenheimer welcomed him and explained what he would like him to do. The laboratory, he said, was staffed and organized to design and make two types of atomic bombs. He stressed, however, that the question of whether and/or how they would be used would be made in Washington. He told Penney that he needed a person who had studied explosive phenomena and blasts in a theoretical way, as well as one who had looked at actual bomb damage. Consequently, Penney spent about half his time working on the scientific phenomena going inwards into the bomb and the other half on scientific phenomena going outwards.[26]

In one of the Tuesday colloquium gatherings, Penney detailed the theory behind the British retaliation bombing. Unaccustomed to such impartial discussions of events like these, several Los Alamos scientists were nonplussed by the contrast between Penney's smiling demeanor and the grim subject under discussion. Because of this, Victor Weisskopf labeled Penney "The Smiling Killer."[27]

Penney played a considerable role at the Trinity test in July of 1945. Kenneth Bainbridge served as overall test director, with John Williams as a principal assistant. Penney acted as a consultant to John Williams on the placement of the instrument lines and inspection of the bunkers and shelters where the people were to be placed at the time of the explosion. He suggested some minor improvements. Penney also worked closely with John Manley, who had the major task of supervising the arrangements to record blast pressures and earth shock for the Trinity test. Penney's assignment was to observe the effect of radiant heating in igniting all structural materials.[28] Once he was standing at the South-10,000 control bunker at Trinity during a dress rehearsal. When somebody spotted him twisting dials and asked what he was doing, Penney jested, "I'm trying to act busy."[29]

Ironically, Penney did not see the Trinity explosion firsthand. He was one of the Los Alamos scientists scheduled to be airborne at a safe distance, part of the tactical maneuvers for the Japanese Mission. The foul weather at the time postponed their take-off and the Trinity shot was fired when they were still in the officers' mess at the Albuquerque Air Base. The decision to ground the bombers was wise, for there was considerable risk that they might have strayed over Ground Zero. As Penney later told Chadwick, "We wouldn't have had a chance."[30]

On July 21, 1945, after the Trinity test results had been somewhat codified, Penney held a Los Alamos seminar. He had studied the blast effects extensively for the target committee and he shared his feelings with the group. Philip Morrison remembered Penney's discussion. "He applied his calculations. He predicted that this [weapon] would reduce a city of three or four hundred thousand people to nothing but a sink for disaster relief, bandages, and hospitals. He made it absolutely clear in numbers. It was reality."[31]

Equally vital was Penney's role as part of the Committee that selected the targets on the Japanese home islands. He was one of five "technical experts" who met to determine this. When the target Committee first met in the Pentagon on April 27, 1944, they knew the seriousness of their discussion. Penney, who had attended scores of meetings, recalled this one as bearing an air of unusual solemnity. Even the normally effusive mathematician John von Neumann was subdued.[32]

General Groves laid down the guidelines for this assignment. The recommendation included four targets.[33] He also told the Committee that their role was only to give "advice." The higher authorities would ultimately make the final decisions. Penney fretted about the nature of the advice he was supposed to give. While he was "fairly sure" that the uranium bomb, nicknamed "Little Boy," would yield from 1 to 5 kilotons of TNT, he had grave doubts about the yield of the (as yet untested) plutonium weapon. It might equal the uranium weapon, but it might also only produce one-tenth of a kiloton.[34] This caused much uncertainty. From above came the directive – Penney still does not know who developed it – that the two Japanese bombs were to be exploded in such a manner that the blast damage should be at a maximum. This meant that they would be exploded in the air, high enough for the fireball *not* to touch the earth. This also meant that there would be no permanent radiation contamination on the ground. Whoever wrote that directive, Penney reflected later, either was lucky or showed great wisdom; he believed the latter to be the case. In fact, the extensive scientific studies that Penney, himself, made probably determined that decision.[35]

Penney was the only British Mission member sent to the island of Tinian as part of the Los Alamos arming team of thirty-seven people.[36] He worked there for three weeks, where his job was to be available if there were any last-minute changes needed in the height at which the bombs were to explode. The results of the Trinity test showed that fuses should be set at the highest of the three settings available. Groves refused to allow any British observer to witness the Hiroshima bombing from the air, but after considerable pressure from London he relented on the Nagasaki bomb.[37] He allowed Penney and an outstanding RAF bomber pilot, Captain Leonard Cheshire, VC, to board the observation plane that accompanied Bock's Car.[38] Thus, Penney was the only member of the British Mission to witness the combat use of the weapons his countrymen had done so much to create. He also visited both Hiroshima and Nagasaki afterwards, so as to measure personally the effects of the bombs.

Penney's expertise was equally in demand for the Crossroads atomic tests of 1946. While surveying the ruins of Hiroshima and Nagasaki, he had noticed that empty gasoline cans had been collapsed there in an inverse ratio from their distance to the explosion. While the Americans were stationing their elaborate measuring devices around the Bikini region, Penney placed several hundred empty Army gasoline containers at strategic areas. Since the first Crossroads bomb missed its target by about a mile, the American instruments proved useless. But Penney's strategically placed gasoline cans provided a fairly accurate estimate of the power of the explosion.[39] After the Bikini test, Penney visited the US for secret discussions on blast effects. He also wrote up a series of classified reports on the matter.

By the time of the passage of the McMahon Act in 1946, Penney probably knew more about the Los Alamos work as a whole than any other member of the British Mission team.[40] The Americans very much wanted him to join them permanently and periodically offered him positions at Los Alamos. While Penney enjoyed the stimulating intellectual atmosphere of Los Alamos, and seemed disappointed that it could not be re-created in Britain, he remained in England. There he faced the most difficult task of his career: overseeing the construction of Britain's atomic bomb. Those who have examined this issue have concluded that Penney's role was central to the enterprise. Historian Lorna Arnold described Penney as "indispensable," while the official historian of the British Atomic Project, Margaret Gowing, termed him the "father" of the British Atomic Bomb. His contemporary, Lord Cherwell, once said that "He is our chief – indeed our only – real expert in the construction of the bomb and I do not know what we should do without him."[41]

From the Maud Report onward, numerous British officials had insisted that Britain needed an independent nuclear deterrent. When a speaker once asked General Montgomery why Britain needed such a weapon, he said, "We must have the bomb, must have it."[42] In late 1946/early 1947, when the decision to proceed was confirmed, Britain felt herself isolated. All the political lessons of the 1930s suggested that military weakness led directly to disaster. The British decision to build a bomb, therefore, did not emerge from any immediate danger. Instead, it arose from the feeling that she needed the bomb to continue to play a major role in foreign affairs.[43] "The discriminative test for a first class power," Penney wrote, "is whether it has made an atomic bomb and we have got to pass the test or suffer a serious loss of prestige both inside this country and internationally."[44]

As head of the Ministry of Supply's Armament Research Department, located first at Fort Halstead in Kent but moved in 1950 to Aldermaston in Berkshire, Penney responded as a patriot to the task at hand.[45] It was his plan (written late 1946) to Lord Portal that Britain produce an atomic weapon under the guise of conventional weapons development. Portal submitted this idea to the Prime Minister and at a small cabinet meeting in January 1947, they gave Penney's plan their approval. Even though Britain claimed the first effective nuclear power plant, at Calder Hall, weapons development remained the number one (albeit secret) priority.[46] This massive effort, in a war-torn country, remained shrouded in secrecy in a Byzantine complex of committees for seven years. The cost eventually reached 140 million pounds. Margaret Gowing termed this "one of the most successfully executed programmes in British scientific and technological history."[47] Even Winston Churchill did not know of the Labor government bomb project until he became Prime Minister again in October 1951.

It is hard to overestimate the problems Penney faced in the early postwar period. Since most of the Los Alamos group returned to academe, British weapons work ever remained short of staff. Some scientists passed up the weapons program because they could never publish their discoveries in the professional journals. Most of the staff came from the defense research establishments. Thus, Penney had to shoulder even more responsibility.

The constant shortage of funds proved equally challenging. While the American weapons laboratories had virtually unlimited moneys, Penney's program limped along on a stringent budget, perhaps one-tenth of that available to Los Alamos, let alone Livermore. An American reporter once marvelled at the austerity of the British program. Lorna Arnold has shown that the success of many a British nuclear test in Australia hinged on the last-minute arrival of some critical component. "We never had the resources we needed," Penney later confessed.[48]

The British public did not know about the independent deterrent until early 1952, when Churchill announced that Britain would test a nuclear weapon during the year.[49] Although Penney had hoped until the last that the test might be staged at the Nevada Test Site near Las Vegas, that never transpired. Both military and political leaders decreed that Britain's test should be conducted independently, and the chosen site became the Monte Bello Islands off the coast of Western Australia. An independent test would show the Americans that Britain was no nuclear satellite; it would also provide ammunition for future political negotiations.

Overseeing the Monte Bello test proved challenging, as the barren and sandy islands had been used only as a stopover for Japanese pearl divers. The Australian government scrambled to provide even basic cartographic information. The site, of course, lacked all the necessities of life. There was no food, water, or timber. The British brought in 400 tons of equipment, including a prefabricated workshop, bulldozers, portable electric power plant, trucks, demolition explosives and armor-plated cable. The Australian Services also provided valuable logistic and physical support.[50]

The bomb was placed in the hold of a frigate and detonated successfully at 8 a.m. on October 3, 1952 (local time). Most of the technical data involved – the explosive lenses, detonators, fuses, etc. – derived from the Los Alamos experience. The weapon was a 20-kiloton plutonium bomb, similar to those detonated at Trinity and Nagasaki, but with several small technical improvements. The test proved the configurations for the Blue Danube warheads, which were then manufactured in quantity for the RAF.[51]

The officially stated objects of the Monte Bello trial were, first, to test Britain's independent nuclear weapon; and, second, to plan for potential Island defense against a surprise detonation from an enemy merchant ship that had tied up in a British seaport.[52] (Because so many British cities were also harbors, this became a major Civil Defense concern of the time.) But the real object was to announce to the world that Britain was still "a great, if not a super, power."[53] The test could not have taken place when it did without William Penney.

When Prime Minister Winston Churchill disclosed the successful outcome of the Monte Bello test to the House of Commons, the MPs hung on every word. When he praised Penney's services, they broke into cheers. Immediately after the test, the Queen awarded Penney a knighthood, which the exhausted scientist first heard over the evening news.[54] A London *Times* reporter praised Penney as "easily the best mind in the world on atom and hydrogen bomb research." He even suggested that Britain's new atomic weapons, the 33rd atomic test at that time (29 American and 3 Soviet) was superior to the almost 30 American bombs.[55] Air Chief Marshal Sir William Dickson, Chief of the Air Staff, coined the phrase, "Dr. Penney's atomic bomb."[56]

Penney himself remained far more modest. In 1953, he said that it would not be fair to compare American and British atomic research, for the Americans had been at it much longer. But the British prominence, he felt, was well deserved, properly integrated, and, on the whole, extremely efficient.[57]

No Americans were present at the Monte Bello test, and their absence was deliberate. The British decision to exclude them was a reaction to the American ban on the exchange of information in the McMahon Act. The sticking point on potential use of the Nevada Test Site had lain with the provision that the British would have had to share the results with the Americans. But as Penney cabled Whitehall from Monte Bello, "We alone will know the power and efficiency of our weapon."[58]

Two months after Monte Bello, Penney requested approval for two other tests (code-named TOTEM). These were moved inland to Emu Field, 300 miles northwest of Woomera, in the desert of Western Australia. These first, land-based British nuclear explosions would later return to haunt the program. Throughout all pre-test negotiations, the British government had repeatedly assured the Australians that the tests would be in no way endanger their population. Three years later, the Australians and British would erect a network of radiation monitoring stations all across the land. But no monitoring stations were present for the two 1953 TOTEM shots, and the first explosion produced the legend of the "Black Mist."

The issue was formally raised in 1980 when a local Aboriginal council advised the Australian Minister for Aboriginal Affairs that some of their people may have been affected by radioactive fall-out. (An anthropologist had heard the story a decade earlier.) Later an Aboriginal, Jim Lester, told of hearing an explosion and watching a black greasy smoke cloud engulf his camp. Many of his people then sickened and, perhaps, some died. Governmental officials denied that the Aboriginals' health problems had been fall-out related, blaming instead the then-current epidemic of measles sweeping the region. The government conceded only that the natives saw the clouds "and misconstrued [them] as harbingers of disaster."[59] But the tale of the "Black Mist" had gone into legend – the legend of irresponsibility – from which no scientific study can ever completely remove it.

As the testing grounds at Emu soon proved inadequate, the British and Australians then signed a ten-year lease on a new site, Maralinga ("Rolling Thunder"), also in the Western Australian desert. Construction began in early 1955 for the scheduled four-shot sequence, BUFFALO, in 1956.

By this time, however, all the nuclear powers had become sensitive to the mounting worldwide criticism against above-ground testing dangers. The Americans' 1954 accidental dusting of the Japanese fishing boat, *Lucky Dragon*, became the symbol of this crusade, but all three nuclear powers shared in the odium. Both British and Australian parliaments and press had begun to voice suspicions. Thus prior to the tests, Penney flew to Australia to hold numerous press conferences to assure everyone that proper safety measures were in place. He also gave several radio and television addresses on the subject. In general, the Australians liked what they heard. Penney's frank, open manner, his sincerity, and his honesty won him many Australian friends.[60] As an additional publicity move, both British Ministers and Australian parliamentarians were invited to observe the tests and were reassured about their safety.

Simultaneously, however, the British weapons program faced yet another challenge. In early 1954, Churchill had decided that the British should develop a hydrogen weapon, and announced this in a February 1955 White Paper. He told Eisenhower about his decision personally. This new assignment increased the burden on Penney and his 200 person staff considerably. While the British had learned some things about hydrogen weapons from the Los Alamos experience, the Aldermaston staff had to derive the weapons from almost whole cloth. Yet within three years, Penney's team had successfully tested the UK's first hydrogen weapon on Christmas Island in the South Pacific.

From 1954 forward, Penney's staff had to (a) stockpile atomic weapons based on the 1952 HURRICANE test; (b) design new, more efficient weapons to reach the 200-bomb goal set for 1957; and (c) begin research and development on a hydrogen bomb. In all of these assignments, the pressure of time proved immense.

Because of increased concern over fall-out dangers, President Eisenhower and Soviet Premier Nikita Khrushchev began test ban negotiations in July of 1955. Penney, of course, never knew when they might reach an agreement. Everyone realized that the British would have had to accept any superpower moratorium on testing. Thus, the whole Aldermaston weapons program reeked with uncertainty from the mid-1950s forward. When a

1958 moratorium was finally agreed upon, America, Russia and Britain all frantically tried to squeeze tests in under the deadline. Weapons development clearly superseded all other concerns of the day.[61]

The Americans had watched the successful British testing program with considerable interest. As has been shown, in 1958 Eisenhower eventually convinced Congress to amend the McMahon Act to allow for exchange of information about atomic warheads. Another provision also allowed for the transfer of fissile materials between the two nations. As historian Timothy J. Botti has argued, the "long wait" of the British had come to an end. By the late 1950s, the US–UK nuclear relationship had become closer and more aligned than at any time since the World War II years.[62]

With this new arrangement, Penney began to breathe a bit easier. He took on other assignments. In June of 1958, Penney served, along with Sir John Cockcroft, as the British representative to the Geneva Conference of Experts to Study the Methods of Detecting Violations of a Possible Agreement on the Suspension of Nuclear Tests.[63] In 1961, he became Deputy Chairman, and three years later Chairman of the UK Atomic Energy Authority. While holding this position, he began to speak out on the benefits of nuclear power. In 1965, he insisted on the advantages in economy and safety of the British advanced gas-cooled reactor and dwelt enthusiastically on the major development effort going into the fast breeder reactors.[64] In 1966, he suggested that the fast reactor was likely to become the cornerstone of the nuclear power system in the future because it would prove immune to any increase in the price of uranium.[65] In general he downplayed the radiation hazards.[66]

Penney attended his first Pugwash Conference in 1960. In 1963, he served as a member of Lord Hailsham's delegation to the test ban talks with the Soviet Union.[67] The next year he addressed the UN on the peaceful uses of nuclear energy. That same year he urged continued fusion research. In 1967, he became Rector of the Imperial College of Science and Technology, one of the largest British colleges with over 3,500 students, where he dealt with student militancy.[68]

Sir William Penney's name returned to the front pages of the newspapers during the early 1980s. Increased concern over radioactive fall-out from the twenty-one British atomic tests in Australia (October 1952 to September 1958) led to the establishment of an Australian Royal Commission in 1984 to investigate the matter. Over 12,000 services personnel had, perhaps, been exposed to fall-out, and it was soon discovered that several "minor trials" at Maralinga (KITTENS, RATS, TIMS, and VIXENS, 1952–63) had permanently contaminated the ground with plutonium.[69] Health concerns over the Aboriginals, who wore little clothing, did not bathe frequently, and slept on the ground were also aired. Emotions ran high in these testimonies, and the Commissioners were often rough on the seventy-five-year-old Penney. His position as senior scientist, plus his open, honest manner, caused him to bear the brunt of the antagonism. In 1987, the Public Record Office furthered the controversy by releasing documents pertaining to the 1957 fire at the Windscale nuclear power plant at Cumbria (now known as Sellafield).

To his dismay, Penney found himself hounded by the press wherever he went. Once his son had physically to fend off reporters who crowded around their vacation spot in Switzerland. As a consequence, Lord Penney broke off contacts with many former acquaintances. He has been reluctant to speak with reporters for several years.

Ernest W. Titterton's career proved equally unique. A native of the Midlands, Titterton had been Marcus Oliphant's first research student, and was teaching at Birmingham University when the war broke out. From Birmingham, he moved into radar research, where he spent the next several years.

In October 1943, he and his wife Peggy (who had been his laboratory assistant at Birmingham) sailed for America. After dodging U-boats, they landed at Newport News. From there they went to Washington, for a brief meeting with Groves, and then to Chicago where they boarded the Santa Fe for Lamy, New Mexico. At Lamy, they were met by Victor Weisskopf and driven to 109 East Palace Avenue in Santa Fe, where Dorothy McKibben prepared their passes. They then were driven up the hill to Los Alamos. On the first evening, they were invited to the Oppenheimers' for a drink and a discussion of the project. Since the Tittertons and Frisch were the first British personnel to arrive, Los Alamos took great interest in them, pressing them constantly with questions.[70]

The Tittertons were assigned a house on bathtub row, right next to the Chadwicks, who arrived two weeks later. Across the road were the Norman Ramseys, the Bainbridges, and the Oppenheimers. As the pathway to Fuller Lodge passed in front of their home, the Tittertons soon recognized everybody by the distinctive nature of their strides. Even the Titterton bathtub was in demand, for the British personnel preferred tubs to the more American showers.

In Birmingham, Titterton had been involved in measuring fission cross sections of U^{235} and U^{238}. Initially he was assigned to Boyce McDaniel's group to continue this on the Harvard cyclotron. Having completed these studies, Titterton began to devote his attention to the complex electronics necessary for both the uranium gun assembly technology and the plutonium implosion technology. Each system required precision timing, with which Titterton had become familiar from his development of spark-gap technology for radar. He also developed one means of instrumentation to observe a sphere under implosion. He was forced to develop special equipment for these experiments, including a pulsed X-ray machine. His group spent most of their time at Anchor Ranch trying to create the proper combination – high velocity outer core and lower velocity inner system – to produce the proper spherical blast wave. Working with Seth Neddermyer, Titterton also did some work with the technology of the gun assembly weapons. These were his contributions at Los Alamos.

His electronics skills were very much in demand for the instrumentation of the Trinity shot. Consequently, Titterton spent many hours in the high desert of central New Mexico. He slept in the Trinity Site dormitories, ate the mediocre food, and watched the nightly, outdoor films. The film program was the brain child of Lieutenant Howard Bush who felt it necessary to keep up the morale of the enlisted men. Titterton felt otherwise. "Hollywood day-by-day," he complained, "was more than the academic mind could take." So, he and his friends sought relief in heckling the films, often to peals of laughter.

Titterton's Trinity Site responsibilities involved the activation of the electronic system that would start all the equipment – especially the numerous cameras that would carry their film through at great speed; and a millisecond later, to generate the electronic pulse that would fire the bomb. He also had to confront two potentially damaging breakdowns. Shortly before the test, a bulldozer broke the main coaxial cable from the shelter to the tower. This necessitated a frantic, last-minute repair. Then, his instruments determined that whenever the weapon assembly on the tower was armed, it fired. This was an error. It should have been only ready to fire, the actual firing coming upon receipt of Titterton's electronic firing signal. Both could have been disastrous.

Titterton and Oppenheimer saw much of each other on the night of July 16. Oppenheimer, whom Titterton described as a "nervous wreck," watched over his shoulder as he inaugurated the electronic firing blip. Neither saw it, however, for just when it was due to appear on the electronic screen, the whole valley lit up behind them. The two men rushed to the door of the control shelter and when the light began to fade, looked at it through smoked glass. Forty seconds later the shock wave hit them. "It was wondrous," he recalled.

After the successful test, the scientists dismantled their equipment and discussed what they had created. When they returned to Los Alamos, they realized "that we had made a most fantastic piece of history, that a new era of energy generation had been opened up, that it could be used for peaceful or warlike purposes, and we would have major problems of controlling this new system, which the world did not yet know about."

When the Crossroads test was approved, Titterton remained in demand. Since the McMahon Act was pending, it took a special agreement to allow him to participate. Titterton prepared all the timing details for the two Crossroads weapons. Working largely with a new team in Washington, DC, he produced the necessary equipment and later deployed it in the Bikini lagoon. On the long voyage to Bikini, he became good friends with the captain of his ship. He also gave a major address to the crew, where he assured them that their virility would not be affected by the fall-out from the Crossroads tests.

After Crossroads, the Tittertons prepared to return to the UK, where he was posted to Harwell, the research arm of the British nuclear establishment. He became part of the General Physics Division there, with the title of Principal Scientific Officer. During his tenure at Harwell, he developed the point of view that he would henceforth maintain: that if anything is possible in nature, it would eventually be produced by men. Thus, new weapons, such as the hydrogen bomb, should be produced, so that Britain could never be blackmailed by other nations. While serving at Harwell, he also advised Penney on the British weapon under construction. He shared with Penney all of the technology he had developed in the States on fast timing, the assembly of lenses, and general timing and fusing techniques needed for implosion. He contributed greatly to the development of Britain's nuclear deterrent. Periodically, Penney tried to entice him into the weapons establishment, but he always declined.

In 1950, Marcus Oliphant approached Titterton with an offer to assume the first Chair in Nuclear Physics at the newly created Australian National University to be built in Canberra. Titterton had mixed feelings about leaving Britain, so he confided his doubts to Chadwick. Chadwick said he should have no misgivings about leaving and prophesied that perhaps he would be able to do more for Britain in Australia than if he stayed at Harwell. As it turned out, this came true.

So, in 1951, the Tittertons moved once again, this time to Canberra. They stayed in a hotel until housing was ready and Titterton then went to work to build up the Australian National University nuclear physics department. "I'm proud to say," he observed later, "that we put nuclear physics on the map in Australia."

When the British decided to test their nuclear weapon in Australia two years later, the Australian government appointed Titterton to supervise the safety of the operation. Thus, he became the Chief Australian Physicist looking after British nuclear weapons in Australia. He continued to monitor all British weapons testing in Australia, both off the Monte Bello Islands and in the interior proving grounds at Emu and Maralinga. When the government founded an Australian Atomic Weapons Test Safety Committee, Titterton was chosen to chair it. He established elaborate measuring instruments, on-site and off-site, involving sticky papers, rainwater samples, sheep bone samples, and even bone analysis from deceased humans. He remained convinced that no Australian citizens, Aboriginal or White, were harmed by the fall-out from weapons tests: "It transpired that all nuclear weapons tests in Australia were carried out completely safely without there being any untoward consequences either to property, fauna or flora, and much more importantly, to human beings." The fall-out stations he established continued to monitor worldwide changes after above-ground testing ceased. In addition, Titterton served on the National Radiation Advisory Commission and the Defense Research Policy Committee, which controlled all weapons research.

In 1985, the Royal Commission investigating British nuclear tests in Australia heard charges that Titterton had been a "special agent" for Britain during those earlier tests. This Titterton stoutly denied. He maintained that the Robert Menzies government had requested that he monitor all tests and declared that he did not play "political games."[71] He also disparaged much of the evidence given to the Commission, insisting that fall-out levels in Australia were always kept well below danger zones – even in light of the new standards. He denied that the earlier tests had been carried out in an unsophisticated way. "They were carried out at a very high level of sophistication," he said, "but it was a level of sophistication applicable to the time and the technology that was available."[72] He remained until his death a bulwark of the Australian nuclear defense establishment.

Although he is not well known to the American public, Danish physicist Niels Bohr was one of the great minds of the twentieth century. He helped develop a new way of thinking that led to the world of quantum physics. His contemporaries placed him in the same category as Archimedes, Sir Isaac Newton, Lord Rutherford and Albert Einstein. When Bohr arrived in Los Alamos under the auspices of the British Mission, he was acknowledged by the scientists as "the father of us all." On his death in 1962, *Time* magazine called him "a man of the century."[73]

Born in 1885 in Copenhagen, Bohr gained fame as a star soccer player and then turned to physics, earning a doctoral degree at Copenhagen in

1911. Afterwards, he journeyed to England to work with both J.J. Thomson at the Cavendish Laboratory in Cambridge and then Ernest (later Lord) Rutherford in Manchester.

Applying Planck's quantum law to the atom in 1913, the 28-year-old Bohr helped formulate what became known as the "Rutherford-Bohr" model of the hydrogen atom, all of which came as a shock to the world of physics. A *Times* reporter wrote that "the ground of physics still quakes from the great upheaval which Bohr initiated and which he had guided ever since with indefatigable patience and penetrating insight." As Frisch said, "With Bohr we gave up the naive concept of reality."[74]

After his sojourn in England, Bohr returned to Denmark to create the Institute for Theoretical Physics in Copenhagen in 1920. There, aided by his charming wife Margrethe, he dominated the field from his mansion "Gamle Carlsberg," given to the nation by the founder of the Carlsberg Brewery. The Institute attracted scientists from all over the world and was compared to Plato's school in ancient Athens.

Bohr also commanded the field through the sheer force of personality. Unlike Einstein, who preferred solitude, Bohr thrived in a crowd. He constantly sought to have students around him. His latest biographer terms him a "radiant human figure."[75] His presence never failed to stimulate discussion, for Bohr always strove for simple answers to complex questions. Ever optimistic, he showed no fear in confronting the probable consequences of momentous breakthroughs. "Every sentence I utter," he said on one occasion, "must be understood not as an affirmation but as a question." "There are things so serious that you can only joke about them," he said on another.[76] His contemporaries held him in genuine reverence. As James Franck wrote, "with such an idea it is always like this: either it is wrong, or Bohr and Einstein have known it for a long time."[77] Even Einstein proved generous with his praise. "I believe," he noted years later, "that without Bohr, we would still today know very little about atomic theory."[78]

By the late 1920s, Bohr's Copenhagen Institute had become a haven of theoretical physics for the entire world. In 1938, it housed Europe's first operational cyclotron. After 1933, the Institute assumed yet another role. It provided a haven for many physicists who had been forced out of Italy, Germany and Russia. Drawing on both Rockefeller Foundation and Carlsberg Foundation funds, Bohr supported many refugee scientists, including Victor Weisskopf, James Franck and Otto Frisch. In addition, Bohr visited the States on an annual basis, in an attempt "to sell his Jews to American Universities."[79]

Since Bohr's mother was Jewish, he was also in danger, and only his devotion to his native Denmark kept him in Copenhagen after the German occupation of 1940. In late September 1942, as Hitler increased his attacks on Danish Jews, Bohr boarded a fishing boat, the *Sea Star*, to flee to Sweden. There he was instrumental in persuading the King to offer unconditional sanctuary to escaping Danish Jews. Documents uncovered at the Nuremberg trials showed that his arrest was imminent. When the Nazis finally raided the Physical Institute in Copenhagen in 1944, they turned it upside down. Later they released a statement saying the building had been occupied because of rumors that Bohr had been working on an atomic bomb (although they found nothing).[80]

The Directorate of Tube Alloys, meanwhile, had kept in close secret contact with Bohr. Finally Chadwick persuaded him to fly to the British Isles. In October of 1943, Bohr departed Sweden in the bomb bay of a single-seat unarmed Mosquito light bomber sent over for that purpose. Only the Mosquito aircraft had the altitude and range to escape the Luftwaffe on the chosen three-hour flight pattern. But this trip proved traumatic for Bohr. He was equipped with ear phones, but they fit badly over his large head, and he did not hear the pilot's instructions to don his oxygen mask. As the plane rose into the thin atmosphere, Bohr passed out. When the pilot received no answer from his famous passenger, he returned at once to a lower altitude and Bohr revived. Bohr's son Aage followed a week later.

When Bohr recovered, Michael W. Perrin took him under his charge. He told Bohr of the great advances that the British and Americans had made in the Tube Alloys/Manhattan Project. Bohr, who had remained unaware of these developments, was dumbfounded. After a month in Britain, Perrin supplied Bohr with a British passport, and Chadwick enlisted him as part of the British Mission. Bohr was officially "Consultant to the British Directorate of Tube Alloys" and Aage, "Junior Scientific Officer."[81] Niels Bohr was seen as the British "trump card" in implementing the Quebec Agreement,[82] and they gladly paid his salary and expenses. Once at Los Alamos, of course, he became "everybody's property."

Eager to gain an overall view of the project, and perhaps even serve as "joint advisor" to both governments, Bohr readily agreed.[83] Thinly disguised as "Nicholas and Jim Baker," Bohr and Aage arrived in New York. There they were met by FBI agents who rushed them into a taxi and then booked them in a hotel, confident that the disguise had been successful. Only as they were registering did an agent notice that "Niels Bohr" was written in large block letters on one of his suitcases. From New York Bohr journeyed to Los Alamos, surrounded by security guards. The manager of Fuller Lodge proved especially deferential to his famous guest. All members of one group at Los Alamos had to sign a document that they would not reveal that the Bohrs were there. Bohr arrived on the Hill surrounded by such mystery that one scientist wondered aloud why he hadn't simply been packed and shipped in a crate.[84]

Upon his arrival, Hans Bethe took him through the laboratory complex and explained the projects in detail. Others sought his advice on numerous issues. As David Hawkins has noted, Bohr arrived at precisely the right moment for the study of the fission implosion process. Some of the most significant experiments were made at his suggestion, and he influenced research on the tamper materials, too. Oppenheimer wrote Groves that Bohr's presence buoyed the morale of all with whom he came in contact.[85] Robert Bacher used Bohr's prestige to resolve a difficult problem. There were two models under consideration for the "initiator" for the plutonium bomb – the device that had to inject neutrons into the plutonium core at precisely the proper moment. Enrico Fermi favored one while the majority of scientists favored the other. Fermi's great prestige made it awkward to vote against him. So, Bacher sought out Bohr in the matter, and he sided with the majority to resolve the issue.[86] Bohr also remained in Los Alamos long enough to file a patent claim.

His lengthy Los Alamos stays were filled with activity. He discussed physics and politics with his prewar junior colleagues. He attended the Pueblo Indian Dances, the social gatherings and further developed his taste for American movies, especially Westerns. He often dined at Fuller Lodge and would stand on the porch looking over at the splendid panorama of the sun reflecting on the Sangre de Cristo mountains. Once, late in the war when he was certain Germany faced ultimate defeat, Bohr jested with William Penney. "Do you know what they are saying in Edinburgh?" "No," Penney replied. "Well, you know that London has fallen and what they are saying in Edinburgh is 'It will be a long war now.'"[87]

Bohr especially enjoyed the mountain scenery. On almost a daily basis he, Aage, and others would take walks through the countryside and discuss scientific questions. Yet Joseph Hirschfelder recalled Bohr as lonesome and troubled during his stay on the Hill. To Hirschfelder, Bohr gave the appearance of carrying the cares of the world upon his shoulders. Philip Morrison also recalled Bohr as the man who thought most deeply about the political aspects of their discovery.[88]

Bohr strongly advocated that the US share nuclear secrets with the Soviets. He made some converts, but the majority of the scientists felt otherwise. When Edwin MacMillan and James Chadwick drove to Compaña Hill for the Trinity test, for example, they discussed the issue. "Why should we?" Chadwick argued. "They don't give us anything."[89]

Bohr's role at Los Alamos was not solely a technical one. Indeed, his chief impact lay in the realm of morale. He smoothed over Army/scientists tensions, continually urging the two groups to work toward a common goal. He also discussed his idea of arms control at every opportunity. A genial man, he soon found himself engaged in intense discussions over the probable aftermath of the Manhattan Project. Ralph Carlisle Smith remembered seeing Bohr and Oppenheimer strolling about the mesa at night, deep in conversation. No one dared come near them as they walked alone. As Oppenheimer later recalled, Bohr "made the enterprise seem hopeful, when many were not free of misgiving."[90]

Although he agreed to the combat use of the weapons, Bohr was also one of the first scientists to anticipate the potential dangers of a US–Soviet nuclear arms race.[91] He launched a one-man crusade to try to forestall such a situation. He was especially insistent that the statesmen not view the atomic bomb as "just another military weapon." Through his contacts with Lord Waverley in England and Judge Felix Frankfurter in the States, Bohr gained access to the top wartime leadership of both nations. "That is why I went to America," he later confessed to Oklahoma physicist J. Rud Nielsen. "They didn't need my help in making the atom bomb."[92]

As Margaret Gowing has shown, however, Bohr's efforts at internationalizing the atom were met with frustration at every turn.[93] Although widely respected within the scientific community, Bohr had notorious difficulty in communicating his ideas. He spoke in a low tone and even his closest friends admitted that his ear lacked the fine sense of English pronunciation. Marcus Oliphant recalled that Bohr used too many words to express his ideas, wandered constantly from the subject at hand, and often dropped his voice so low that he became inaudible. As Rudolf Peierls noted, Bohr proved better at talking than listening.[94]

The tales of Bohr's problems of communication have become legendary. On one occasion he spoke until almost midnight. On another, he gave a talk to the Kapitza Club at Cambridge on the uncertainty principle. After a rambling, two-hour presentation, Lord Rutherford had to halt his guest by jesting that Bohr's sense of time was as uncertain as the principle he was discussing.[95]

In the early 1930s, John Manley heard Bohr at the University of Michigan but found that the low, almost inaudible delivery obscured the message. In the late 1930s, Bohr gave a talk *against* prejudice. After the address a member of the audience asked why he hated the Swedes so.[96] Ralph Carlisle Smith told of a similar incident at a Los Alamos colloquium, when Bohr presented a slide discussion of a scientific problem. Large segments of the audience disappeared with regularity every time Bohr darkened the room to show slides.[97] After another lengthy Bohr presentation, a scientist leaned over to a colleague and confessed that he simply couldn't understand Bohr's Danish. As it turned out, Bohr had been speaking English. Donald Marshall once heard Bohr discourse at length on "the whiz of the devils," only to discover that he meant "the widths of the levels."[98] General Groves once sought out Bohr's understanding of European science and spent several hours in conversation on the train to Los Alamos. When Oppenheimer ran into Groves the next day, Groves seemed somewhat out of sorts. When Oppenheimer inquired why, Groves smiled, and said that he had been listening to Bohr. This inability to communicate proved a lifelong handicap, one of which Bohr was only dimly aware. As Oppenheimer noted, it was easy "for even wise men not to know what Bohr was talking about."[99]

Bohr's written prose was not much clearer. He revised his scientific papers over and over again, adding clauses and reservations, so that his earlier drafts were often easier to understand than his later ones. He tended to revise his articles up until the last moment, striving for the exact meaning in every sentence. His colleagues who wrote joint papers with him experienced this frustration first hand.[100] His collected "philosophical writings" also make for difficult reading.

In April, 1944, Bohr returned from the States to England. He pressed his case for internationalization of the potential weapon with vigor. For him it had become "a matter of urgent anxiety."[101] Thus, he shared his ideas with Sir John Anderson and Lord Cherwell. He finally gained a conference with Churchill in mid-May of 1944.

Bohr's inability to communicate was most dramatically highlighted during this famous meeting with Winston Churchill on May 16, 1944. Preoccupied with the imminent invasion of the Continent and overly sensitive about the terms of the Quebec Agreement, Churchill only reluctantly agreed to see "The Great Dane," as he was known in Whitehall. Their half-hour discussion went nowhere. As the disappointed scientist prepared to depart, Bohr asked the Prime Minister if he could write him a letter. Churchill replied: "It will be an honour for me to receive a letter from you . . . but not about politics."[102] Later Churchill told Lord Cherwell, who had helped arrange the meeting, "I did not like the man when you showed him to me, with his hair all over his head."[103] A disappointed Bohr ran into R.V. Jones shortly after his encounter with Churchill and Jones asked him how it went. Bohr sadly replied, "It was terrible. He scolded us like two schoolboys."[104] While it is perhaps an overstatement to suggest that the postwar nuclear arms race hinged on Niels Bohr's inability to make himself understood, one could make a good case along these lines.[105]

Three months later Bohr returned to the States for an hour and a half meeting with President Roosevelt. This one seemed to go a bit more smoothly. Bohr departed feeling that the President was on his side. But when Roosevelt and Churchill met on September 18, 1944, Churchill again regained the upper hand. In their joint *aide-mémoire*, the Prime Minister convinced the President that Bohr's ideas about internationalizing the atom formed a potential menace to international security.[106] The FBI began to watch Bohr to make certain he did not try to share atomic secrets on his own. Churchill obviously did not understand the man they were dealing with. A scientist of great personal integrity, Bohr made no attempt to contact the Soviets. Niels Bohr was no Klaus Fuchs.

Bohr's solution to the problems of world peace in an atomic world was both simple and complex. He passionately believed that the only chance of safety for mankind would be to make an open offer to share the knowledge of this discovery with all the nations of the world. In return, these nations would agree to surrender further claims to national secrets about science. An international agency would be largely composed of an international cadre of scientists.

In retrospect, Bohr's ideas about an "Open World" of free exchange of scientific atomic data with all nations seem a little naive. Historian Donald Cameron Watt called Bohr's scheme a "flight into the higher mysticism" away from the unpleasant and unacceptable world of politics. Bohr's ideas flowed naturally from two formative areas of influence on his life: the late nineteenth-century growth of European internationalism and the spread of upper-middle-class professionalism, especially in the hard sciences. During the 1920s and 1930s, the world of physics knew no national boundaries. In his essays and speeches, Bohr frequently praised the "unique cooperation" of a generation of international scientists who helped forge "a vast new domain of knowledge."[107] The international aspect of theoretical physics was nowhere better illustrated than in Bohr's own Copenhagen Institute, with its visiting scientists from all over the world.

But this atmosphere would not long survive World War II. Science was rapidly dividing into ideological camps, and the world of technology – which has never been shared by anyone – quickly became the most closely guarded of national secrets. While Western Europe might eventually move toward political internationalism of sorts, the problems of international terrorism introduced yet a new factor into the equation.

Moreover, in 1944–5 there existed no real international order to carry out these plans. The League of Nations lay in shambles. The United Nations remained an unproved commodity. The wartime alliance of Russia/Great Britain/the US was a marriage of convenience at best. Any such "triple alliance," especially one bound by great historical antagonisms, could hardly be expected to survive. Niels Bohr, unfortunately "was as ignorant of politics as Churchill of nuclear physics, [but] unlike Churchill he was unaware of his own ignorance."[108]

But if Bohr failed to convince either Churchill or Roosevelt of his plans to internationalize the atom, he did have considerable impact among the scientists at Los Alamos. He found an early ally in Oppenheimer who helped clarify his ideas for popular consumption. He also convinced British scientist Sir John Anderson (Lord Waverley). Philip Morrison later noted that his own postwar political stance toward the bomb and its international meaning were all greatly influenced by his contact with Bohr.[109] Joseph Rotblat confessed that he, too, owed much to Bohr's stimulating conversations. In a sense, then, both the initial American attempts to establish international control over atomic weapons and the Pugwash Conferences owe their existence to Niels Bohr's Los Alamos ideas. So, too, do the current United Nations and European international atomic committees.

In June of 1945, Bohr left the States permanently for England. He remained there until the end of the war, when he returned again to his beloved Copenhagen. He later served as Chairman of the Danish Atomic Energy Commission and helped establish the national nuclear power plant at Riso. He also became active in the United Nations Atoms for Peace program. However, he had no further connections with any military application of atomic energy. In private, he agonized over the fact that the Allies had used the bomb in anger.[110]

For several years after the cessation of hostilities, Bohr retained a low profile on political issues. In 1946, he wrote a "foreword" to the Federation of Atomic Scientists booklet, *One World or None*. That same year he submitted a lengthy memo to the British Advisory Committee on Atomic Energy. In both 1946 and 1948 he utilized his visits to the States for scientific conferences to approach American statesmen about his ideas. In May of 1948, he submitted a memorandum to the Secretary of State that served as the basis for extensive further discussions a month later. None of these conversations received much publicity.

Then in June of 1950, Bohr once again broke into the spotlight. He wrote a lengthy "Open Letter to the United Nations" setting forth his ideas for a peaceful postwar world. *The Bulletin of the Atomic Scientists* sent copies of Bohr's letter to any reader who would send them a four-cent stamp.[111]

Dismayed by the hardening of postwar attitudes, Bohr once again pleaded for the creation of an "open world" in the exchange of scientific, especially atomic, information.[112] He argued that the scientific and technological progress had tied all nations inseparably together. A scientific openness with "free access to all information of importance for the interrelations between nations" was necessary for both the reduction of Cold War tensions and as a way to ease international understanding and cooperation.[113] Yet Bohr was also a realist. He often posed the question: how could Western Europe have remained free after World War II if the Americans had not had the bomb?[114]

Bohr's "Open Letter to the United Nations" sparked considerable discussion throughout all of western Europe. It aroused special comment in Scandinavia, and brought Bohr back briefly into the political limelight. Danish Communist Party members challenged him publicly on his ideas. Bohr himself became somewhat alarmed about the interpretations being given to his letter. He stated that too much emphasis had been placed on his call for openness from the Americans. He emphasized that Russian secrecy had to be lifted, too.[115] The political reaction to this letter, and the failure of any government to do anything constructive, disappointed him further. In 1955 he organized the first Atoms for Peace Conference in France. He was also instrumental in establishing the Centre Européan pour la Recherche Nucléaire (CERN) in Geneva.

Although Bohr worried constantly about nuclear issues, he refrained from further public statements during the twelve years he had remaining to him.[116] His death in 1962 marked the passing of a generation. Niels Bohr's fervent pleas in this regard had one main goal: to obviate an American–Soviet arms race that he predicted would be disastrous to all involved.

6 The Strange Tale of Klaus Fuchs

Emil Julius Klaus Fuchs has emerged as the most notorious member of the British Mission to Los Alamos. Unnoticed at the time, his name became a household word in 1950 when he was exposed as a Soviet agent. Thanks to the recent declassification of FBI documents on the case, the public has more data on Fuchs than on any of the other "Atomic spies." In spite of this information, however, Klaus Fuchs remains almost as much an enigma today as he was to his colleagues at Los Alamos.[1]

Unlike several of the other Eastern Bloc spies, Fuchs retained a low profile after his release from prison and flight to East Germany. He never appeared on Soviet television, as did Ruth Kuczynski Beurton ("Red Sonia"), the infamous Kim Philby, or recent CIA defector Edward Lee Howard. Fuchs never wrote a memoir along the lines of Philby or Red Sonia. Unlike Italian defector Bruno Pontecorvo, he was never allowed by his superiors to visit the West again. In fact, Fuchs once had to deny rumors that the East German government had confined him to prison.[2]

Fuchs was not exposed as a Soviet espionage agent until four years after he left Los Alamos, when he was a fixture at Harwell. With a little better luck, he might have retired as head of the theoretical physics division of Harwell, laden with honors. That he did not was largely due to the efforts of FBI Agent Robert Lamphere.

In the fall of 1947, Lamphere became supervisor of the Espionage section of FBI headquarters in Washington, DC. Among other items, he inherited a safe filled with numerous KGB messages from the 1944–5 period. These had all been beamed from the Soviet Consulate in New York to KGB headquarters in Moscow, but only a fraction had been decoded. Lamphere assigned a top level cryptographer, Meredith Gardner, to the case and Gardner soon made several breakthroughs. In 1948, the FBI discovered that somebody had supplied the Soviet Consulate with a top-secret scientific report on the gaseous diffusion method of producing U^{235}, as well as other Manhattan Project data. Soon they became convinced that the KGB had had an agent within the British Mission to the Manhattan Project.[3] Klaus Fuchs had written the scientific reports summarized in the intercepted messages, but that, by itself, did not necessarily indicate guilt. Initially, the FBI suspected all four British Mission members who had been stationed in New York City: Christopher Frank Kearton, Fuchs, Tony Skyrme, and Rudolf Peierls.[4]

Rudolf Peierls was suspected because his eminence in British scientific circles allowed him access to many secret documents. In addition, he was a German refugee and his wife, Eugenia, was Russian by birth. The initial suspicion evaporated when the FBI could produce no firm evidence that Peierls had engaged in espionage. But rumors to the contrary continued to surface over the years. In 1979, Peierls brought a suit against British publisher Hamish Hamilton for the book, *The British Connection: Russia's Manipulation of British Individuals and Institutions*. Written by *The Sunday Times* reporter Richard Deacon, *The British Connection* claimed that Peierls had been responsible for massive leakage of restricted data. When rumors to this effect reached the Oxford academic community, they caused Peierls much embarrassment. Deacon also claimed that Peierls had died. "I am still alive," Peierls told a *Daily Telegraph* reporter. "The author's statement is about as accurate as the rest of the book." Eventually, Peierls' suit proved successful. The author and publisher paid "very substantial damages" and removed the offensive passages from the book.[5]

Tony Skyrme fell under suspicion for less obvious reasons. A 1943 Cambridge graduate in mathematics, Skyrme had previously worked under Peierls at the University of Birmingham. In February 1944, Skyrme joined the British Mission team in New York City. Fellow British Mission member Christopher Frank Kearton proved the chief source of anti-Skyrme data. Kearton told FBI investigators that if there were a disloyal member in the British delegation to the United States, it would most likely be Skyrme. Kearton recalled Skyrme as a man of strong convictions, and one very likely to have developed radical social contacts in New York City. He also recalled that Skyrme had become involved in several minor scrapes while he was in the States and was "rather odd."[6] These accusations were also without foundation.

By the fall of 1949, the FBI had become convinced that the primary leak was Fuchs. They informed British security, MI5, to that effect, and the British officials began to devise a plan to expose him, without disclosing the breaking of the Soviet codes. Fortunately an opening appeared when Fuchs's aged father, Emil Fuchs, announced that he would shortly move from the western zone of Germany to accept a new academic post at the University of Leipzig in East Germany.

The potential for embarrassment or blackmail in the elder Fuchs's new situation seemed obvious. Consequently, Klaus Fuchs approached his friend, Wing-Commander Henry Arnold, who served as MI5's security officer at Harwell, and asked if he should resign. Arnold, who had long harbored suspicions about Fuchs, listened carefully but offered no advice. Arnold utilized the incident, however, to bring in master investigator William James Skardon from London to further query Fuchs. For several weeks Skardon met with Fuchs on this issue. He carefully chose the moment to confront him with the charge of espionage. At first, Fuchs seemed surprised and denied everything. But on January 13, 1950, he confessed. Judging from his confession, Fuchs seemed to believe that his father's impending move left him with only two alternatives: he could conceal his past and leave Harwell. Or, he could tell the truth, clear the air, and remain at Harwell.[7] Although the idea of suicide crossed his mind, Fuchs showed no realization of how his espionage would be viewed by the British authorities or the British public.

The government moved as swiftly as possible. Fuchs was arrested on February 2, charged the next day, given a hearing on February 10, and brought to trial on March 1, the earliest possible date. The charges against him were that he passed secret data at four different times: in Birmingham, England, in New York City, in Boston, Massachusetts, and in Berkshire. Instead of being charged with high treason (the British were not at war with the Soviet Union) Fuchs was charged with violation of the Official Secrets Act. All knowledge of the FBI code breaking was kept secret, and the public was told that the government had no independent evidence other than Fuchs's own confesssion.

British officials were reluctant to let the FBI interview Fuchs in prison. British tabloids suggested that the FBI would use "third degree" methods and that they planned to extradite him to the States for trial. Actually, because of the secrecy surrounding the broken Soviet codes, the FBI had no desire to bring him to the States.[8] Eventually the British allowed the FBI to interview Fuchs, and Robert Lamphere's fifty-page summary of this discussion is a gold mine of information.[9]

The trial was brief and to the point. The prosecutor presented Fuchs as a classic case in English literature, that of Dr Jekyll and Mr Hyde. Fuchs himself concurred. He declared that he had divided his mind into two separate compartments, which he described as a "controlled schizophrenia."[10]

Under the glare of great publicity, Fuchs pleaded guilty to all four counts of the indictment. He even thanked the court and all concerned for a fair trial. The court stated that the crime Fuchs was charged with differed only thinly from high treason. The judge sentenced him to the maximum penalty – fourteen years.[11] (He would go first to Wormwood Scrubs and later to Wakefield gaol.) The official Russian news agency TASS denied any knowledge of Fuchs.

When the public learned that "Britain's third ranking atomic scientist" had been convicted for treason, they were outraged.[12] One newspaper termed him "the cleverest spy who ever operated in Britain."[13] The judge who passed sentence called Fuchs "one of the most dangerous men the country could have on its shores." Afterwards, the British press worried at length if the nation could continue to accept political and religious refugees. One of the tabloids splashed front-page photos of other émigré scientists – Frisch, Peierls, Born, and others – and declared (erroneously) that these émigrés had given their passports to authorities to show their loyalty to the nation.[14] British security received an enormous amount of criticism for its failure to spot Fuchs earlier, and the Prime Minister had to defend MI5 on the floor of the Commons.

The treason of Klaus Fuchs proved very different from the rash of 1980s espionage cases. Most of these spies traded secrets to the Soviets for sex or money, or to satisfy personal pique. Fuchs had purely ideological motives. Born into a generation that saw Soviet Communism as the only bulwark against fascism, Fuchs never varied from his early student radical position. Over the years, he remained an ideological purist.

As biographers Robert Williams and Norman Moss have shown, Fuchs was hardly a "professional spy." Rather, he was a "professional scientist" who had been encouraged to join the Tube Alloys program because the British needed his mathematical physics skills.[15] Fuchs became a spy out of conscience. He proved important only because the discoveries in atomic physics were the most startling scientific discoveries in the world at that time.[16] Had Fuchs been a biologist, geologist, or astronomer, no one would have ever heard of him. Moreover, one looks in vain for any press mention that Fuchs was also assisting the British in their secret program to build their own atomic bomb.[17] The Fuchs trial concealed as much as it revealed.

The Fuchs case, however, had major consequences on both sides of the Atlantic. On one level, the exposure of Fuchs doomed the delicate negotiations underway to integrate more tightly the American, British, and Canadian nuclear defense plans. The British and Americans were close to an agreement, but the barrage of adverse publicity and increasing American suspicion of British security measures doomed the proposals. The British–American nuclear relationship remained strained for several more years.[18]

In the wake of the trial, the FBI launched one of the greatest manhunts of the century. Their goal was to discover Fuchs's secret contact and (perhaps) other co-conspirators as well. Overall, the FBI investigated perhaps 1,200 people, including scores of scientists. Among these were Rudolf Peierls, Carson Mark, and George Placzeck of the British Mission. The manhunt showed FBI scientific detective work at its best. Eventually, on May 22, 1950, the FBI found Fuchs's American contact in Harry Gold, a Philadelphia chemist. Gold, in turn, led the FBI to David Greenglass, a Los Alamos machinist, and he accused his sister, Ethel Rosenberg, and her husband, Julius, of espionage.

The investigation of Fuchs led to eight arrests, including Gold, Greenglass, Julius and Ethel Rosenberg, Morton Sobell, Abraham Brothman, and Miriam Moskowitz. In all, over forty-five espionage cases were opened because of this.[19] Greenglass made a deal with the FBI for a reprieve; Gold received an 80-year prison sentence; and, in 1953, the Rosenbergs were executed in the *cause célèbre* of the decade. From his cell in Wakefield gaol, Fuchs followed the Rosenberg case with interest. Once he remarked that he was fortunate not to have been tried in America; otherwise, he surely would have met the same fate.[20]

The American public was as outraged over Fuchs's espionage as the British. A spokesman at the Senate Hearings on Soviet Espionage credited Fuchs with accomplishing "greater damage than any other spy, not only in the history of the United States but in the history of nations." A University of Illinois scientist wrote President Truman that Fuchs was "the most harmful traitor of them all."[21] FBI Director J. Edgar Hoover wrote an article for *Reader's Digest* on the case entitled "The Crime of the Century." Eventually that phrase came to be associated with the Rosenbergs, but Hoover intended it chiefly to refer to Fuchs.[22]

Identical public outrage, however, produced quite different political consequences. British law was clear on the matter. Because Great Britain was not at war with the country to which Fuchs had passed his information, the crime was not treason. So, Fuchs received a fourteen-year sentence, not execution. In addition, British anger over Fuchs dissipated rather quickly. It never escalated into the McCarthyism of the United States. In 1949, Rudolf Peierls observed that, unlike the Americans, the British never really worried about the dangerous effects of scientists who held "subversive" views of one type or another.[23] Perhaps the British were less paranoid about the Russians; or perhaps they had longer acquaintance with espionage than the Americans.

The furore surrounding the Fuchs case proved an integral component of the anti-communist hysteria of the 1950s. In early 1950, Congressman J. Glenn Beall (Republican, Missouri) argued over national radio that every American contact of Fuchs should be investigated and "brought to justice." He predicted that more Russian agents would be exposed from this case than in the infamous Canadian spy investigation of 1946.[24] However, the spy cases faded with the execution of the Rosenbergs, who refused to name other names. The current interpretation of this case suggests that Hoover did not want them executed, and only reluctantly gave his consent. Hoover hoped that the Rosenbergs would list other conspirators, and that the case would go even further.[25]

The timing of the Fuchs case also had considerable popular repercussion. The American public learned about Fuchs only three days after Truman's announcement that the government had decided to pursue a hydrogen bomb program. While Fuchs had *not* disclosed much of value on the H-bomb, his treason certainly helped sway public support for this latest development in nuclear weaponry. The diary of AEC Commissioner Gordon Dean says little about Fuchs, but we know that the Commission spent hours of secret discussion trying to assess the exact nature of the material he passed. In the Spring of 1952, Hans Bethe wrote a top-secret, detailed memorandum on what Fuchs might have passed on regarding H-bomb research.[26]

The strange tale of Klaus Fuchs continues to fascinate. Born in Russelsheim, Germany, near Frankfurt, on December 29, 1911, Fuchs grew up in a religious family. His father, Dr Emil Fuchs, served as a Lutheran Professor of Theology, and later became an active Quaker. The elder Fuchs achieved an international reputation in Quaker circles for his writings and lectures. Eventually he was sentenced to a concentration camp for his opposition to the Nazi regime. Fuchs's older sister, Elizabeth, killed herself because of Nazi political harrassment.

Klaus Fuchs was active in communist circles when he was a student at the University of Kiel, and he fled to England only after the Nazi takeover. Arriving in England in 1933, he studied under Nevill Mott and eventually earned a PhD at Bristol University, where he made little attempt to hide his communist leanings. Later he moved to Edinburgh where he received an ScD in 1936 under the tutelage of fellow refugee, Max Born. In May of 1940, Fuchs was classified as an "enemy alien" and interned, first on the Isle of Man and then in a detention camp in Canada. He arrived back in Britain on December 17, 1940. In June, 1942, he became a naturalized subject. Rudolf Peierls was so impressed with Fuchs's work that he invited him to join the Tube Alloys Project. Fuchs actually lived at the Peierls household for an eighteen-month period. As Alice Kimball Smith noted ruefully, "The Peierls were awfully good to him."[27]

Once Fuchs learned of the true nature of his work, he decided on his own to inform the Soviets. He sought out an English Communist Party member who soon put him in contact with a courier. Fuchs made his decision because he felt that the Allied Powers had deliberately planned to allow

Germany and Russia to bleed each other to death. At first he supplied the couriers with only his own work, but when they requested more details, he gave them other materials as well. He did this on at least six occasions. When he informed them that he was about to be posted to New York, they made arrangements for a new stateside contact.

Sailing on the *Andes*, Fuchs arrived in the States at Norfolk on December 3, 1943. He was immediately sent to New York City as an employee of the British Department of Scientific and Industrial Research, which had offices at 43 Exchange Place and 37 Wall Street. For several months Fuchs roomed at the Taft Hotel and Barbizon Plaza Hotel, and during that time he met his American contact Harry Gold ("Raymond").

The chief task of the British Mission scientists in New York was to help the Kellex Corporation by analyzing certain theoretical studies regarding the gaseous diffusion process of uranium isotope separation. Their findings were summarized in a series of reports called the MSN series. They shared these reports with the Columbia research group. The reports were vital to the plant design under way at Oak Ridge, and Groves had told the Kellex officials that they could discuss anything with the British delegation, so long as both sides were engaged in work on this subject. The general warned Kellex only that they should not go beyond the matters under discussion, especially long-range atomic plans.[28]

The British team in New York worked hard to perfect the gaseous diffusion process. At first they engaged primarily in preliminary discussions, carefully reviewing the proposed K-25 plant design. (K-25 received its code name from the location on a planner's grid.) From December 1943 through May 1944, the team provided assistance in solving various theoretical problems. Finally, from July until departure, they worked on problems pertaining to the scraper cold traps. Valuable though their calculations were, none of the British scientists ever visited the Oak Ridge gaseous diffusion facilities in person (although other British Mission members from Berkeley later visited Oak Ridge's electromagnetic separation plant).

The MSN series of reports summarized the British contributions on the question of gaseous diffusion. Fuchs wrote over thirteen of these papers, while Peierls and Skyrme wrote several each.[29] The reports proved crucial to the successful separation of U^{235} and U^{238}, both in Oak Ridge and later in the Soviet Union. After they had finished that assignment, the three were posted to Los Alamos.

Once Fuchs moved to Los Alamos, contacts with his courier proved more difficult. During his nineteen-month stay, Fuchs met Gold only four more times, twice while he visited his sister in Boston and twice in Santa Fe. For vacation Fuchs travelled to see his sister, Krystal, who lived in the

Boston area. During his visits, he saw his first American jazz performance, and also learned about his sister's increasing marital and mental problems. He utilized each visit to turn over written materials to Gold.

The first Santa Fe meeting occurred on a Saturday in June of 1945. The two men met on Alameda Street on the north bank of the river, between Castillo and Delgado streets. Here Fuchs passed over a longhand, personal sketch of the workings of the implosion bomb. Following orders from his Soviet superior, Gold offered Fuchs about $1,500 at this meeting, but Fuchs refused. Upon departure they agreed to meet again in September 1945.

When the September appointment arrived, Fuchs picked Gold up in his car at Washington Avenue, just below Kearney Street, in Santa Fe, and proceeded out the Bishop's Lodge Road. Fuchs parked the car, and they had a long talk. He told Gold of the excitement of the Trinity test and how the nation had suddenly begun to treat the scientists as heroes. After handing over some materials, the two parted, and Fuchs never saw Gold again.

Fuchs's stay in Los Alamos was nothing extraordinary. He arrived on the Hill on August 14, 1944 and lived in several places: the Big House, at Room 17, Dorm T-102, and Room 5, Dorm T-102. During his stay he kept largely to the company of the other British scientists, especially Peierls. Genia Peierls – "Mother Peierls" to many of the men – tried to look out for him because of his apparent loneliness.[30] Both Fuchs and Peierls were assigned to Hans Bethe's Theoretical Division, and both did valuable calculations on blast-related issues. Their calculations played a vital role in perfecting the implosion process. As Norris Bradbury later observed, Fuchs worked very hard for the United States. His work so impressed Bradbury that when he took over as Director of Los Alamos, he asked Fuchs to delay his final departure by four months.[31] When Fuchs finally departed, he carried with him most of the vital information about the workings of the bomb.[32]

After Fuchs's treason was exposed, rumors abounded that Los Alamos had assigned a team of mathematicians to re-do all of his earlier calculations, to see if he had attempted sabotage as well as treason. Carson Mark, who headed the Los Alamos theoretical division then, denied this. Fuchs worked as hard for the Allies as he did for the Soviets, Mark observed. He would have been mortified if he had made an error in calculation.[33] Hans Bethe also praised Fuchs's scientific work. "He was one of the most valuable men in my division," Bethe said, "one of the best theoretical physicists we had."[34] Another colleague recalled Fuchs as an especially lucid lecturer at Los Alamos. Even today, Fuchs's report in the LA 1020 Series on scaling for blast waves is considered a classic of that field.[35]

When Fuchs was exposed as a spy, there was universal shock among his former colleagues, both in England and at Los Alamos. None of his mentors – Born, Peierls, or Bethe – had the slightest inclination of his double life. His friends were equally dumbfounded.

At 5'9", 150 lbs, with glasses, Fuchs was a quiet man but no recluse during his Los Alamos days. Going by "Karl" rather than "Klaus," he dated Evelyn Kline, a grade school teacher, taking her to several dances and a night club. He also dated another grade school teacher, Jean Parker. He could be found dining at the homes of his fellow scientists on the average of twice a week. He enjoyed babysitting his colleagues' children, so as to free the parents to attend parties. Fuchs danced, drank considerably, skied, and often climbed the nearby mountains. He also enjoyed hiking the numerous back trails of Bandelier National Park. At times he even showed a sense of humor. After an eight-hour hike to view a pair of Anasazi stone lions in the Bandelier backcountry, Fuchs joked that "I have not seen Chicago, but I have seen the Stone Lions."[36]

During his stay on the Hill, Fuchs purchased a secondhand Buick and took several tours of the surrounding countryside. Immediately after the war, he, Mici Teller and both Peierls took a vacation to Mexico. Before he left he jested with Tony Skyrme that he had a secret rendezvous with a beautiful lady there. On the way, they had car trouble and the group spent two days in Marfa, Texas, described by Fuchs as "a Texas metropolis of 3,000 inhabitants."[37] From Mexico he sent his nephew, Steven Heineman, a Mexican version of an erector set. One of his biographers feels he was happier in Los Alamos than he had ever been in his life.[38]

When his former colleagues tried to recall the Klaus Fuchs they once knew, they often stumbled. The enlisted men of the Special Engineer Detachment (SEDs) could recollect him, largely because they turned his surname into an American vulgarism. Ralph Carlisle Smith remembered the exact details of the man, mostly because Fuchs wore the same brown sport coat and slacks every day. Others who dealt with Fuchs on a regular basis, however, could remember almost nothing about him. Skyrme shared an office with Fuchs but saw him rarely in a social setting. One colleague recalled him as "a charming fellow" while another termed him "a colorless, disembodied and methodical brain." Genia Peierls described him as "penny-in-the-slot" Fuchs because a person had to offer a sentence to him to receive one in return. Edward Teller felt the same way: "In talking, his spontaneous emission is very low but his induced emission is quite satisfactory."[39] Elfriede Segré called him "Poverino," the pitiful one. One Los Alamos wife described him as a very quiet, rather sweet, reticent little guy; another, as a mild, unobtrusive pleasant man who never talked politics. Both Laura

Fermi and Dorothy McKibben remembered Fuchs's politeness and refined manners. Ruth Marshak developed a strange fondness for him because he seemed so lonely. The New York secretaries remembered him only as a shy, efficient, and businesslike scientist. His landlady in Abingdon told reporters that while he understood Einstein, he could not tie a bow tie or run a car.[40]

Richard Feynman probably knew him as well as anyone at Los Alamos for they both lived in T-102 dormitory. On an almost nightly basis they discussed politics, Lab security, the free flow of scientific information between nations, and a wide variety of other issues. Yet four years later, Feynman could recall no strong views that Fuchs held on any item. In the intense political discussion in Los Alamos from 1944–5, many scientists voiced the opinion that it was unfair to exclude the Russian Allies from the Manhattan Project. The scientists realized full well the nature of the new and terrible weapon on which they were working. The consensus was that effective postwar international control of weapons depended on Russian participation. A radical few voiced the opinion that if Washington failed to realize this, the safety of humanity demanded that they tell the Russians by themselves. Physicist Martin Deutsch noted that virtually every scientist participated in these discussions, at least in part. Only Fuchs declined to comment on these issues.[41]

Since Richard Feynman's wife, Arlene, was dying of tuberculosis in Albuquerque, Fuchs showed special concern for him. He lent him his old Buick so that Feynman could drive to Albuquerque on a moment's notice. After Arlene's death, when Feynman refused to grieve openly, Fuchs was even more considerate. He made certain that Feynman attended every possible evening gathering and watched over him closely for several days.[42]

Once Feynman and Fuchs had bantered the question as to which one would be the most likely candidate to be a spy. They decided that Feynman would be the more likely spy suspect because of his frequent trips to Albuquerque to visit his ill wife. In early 1946, Feynman asked Fuchs why, in view of the austerity of British living conditions, he did not remain in the United States. Fuchs replied that he felt he had an obligation to continue his scientific work for Great Britain.[43] Indeed, Fuchs returned to the States only once after he departed Los Alamos: for a 1947 British-American Conference on declassification.

In retrospect it is clear that nobody at Los Alamos really knew Klaus Fuchs. Peggy Titterton complained that "you just couldn't get on with him at all. He was not forthcoming."[44] Martin and Suzanne Deutsch often entertained Fuchs as a house guest but confessed that they only knew "the shell or hull of the man."[45] Others remembered Fuchs as "queer" or "not an ordinary person," but mild eccentricity was not at all unusual in wartime Los Alamos.[46] "I have to admit a complete failure to understand Mr. Fuchs," Norris Bradbury said.[47]

In the early 1950s, American officials spent a great deal of time discussing just what material Fuchs had passed on to the Soviets. In 1950, D.J. Littler was asked to survey the Klaus Fuchs documents at Harwell. Littler discovered, to his dismay, that Fuchs had systematically accumulated an enormous amount of detail on both his as well as other people's work at Los Alamos. Everyone assumed that all of this ended in Soviet hands. Fuchs passed on all he knew about the gaseous diffusion method of isotope separation at Oak Ridge. He described the details of the implosion process – the most difficult aspect of weapons production. In addition he revealed all the discussions over the hydrogen bomb, data on which he continued to supply as late as 1948.[48] (The "Classical Super" process then under discussion, however, proved unworkable.)[49] The only items he failed to supply were the specific details of engineering.

Groves always insisted that it was the lack of compartmentalization at Los Alamos that had allowed Fuchs to gain his considerable knowledge of the Project. Oppenheimer, however, felt otherwise. "Even if Fuchs had been infinitely compartmentalized," Oppenheimer observed, "what was inside his compartment [the Theoretical Division] would have done the damage."[50] Hans Bethe put it this way: "[Fuchs] told the Russians exactly how to assemble the bomb; how to use implosion; how the explosive and nuclear material was arranged; how to calculate the yield of the bomb and the neutron diffusion, and he certainly told them about the critical mass – the size of the melon – which is the most important for knowing the size of the reactor you must build to produce the material."[51] He knew everything the Americans did and he delivered that information to the Soviets.[52]

While the Soviet Union has yet to publish its official version of a Smyth Report, recent scholarship has made our knowledge of their early program considerably clearer. Because of their hopes that science might promote economic growth, the early Bolshevik leaders of the 1920s did not impose ideological restraints upon the members of the Russian Academy of Sciences. In 1921, for example, Russian physicist Peter Kapitza journeyed to the Cavendish Laboratory at Cambridge to work with Lord Rutherford. Brilliant and charismatic, Kapitza immediately organized the "Kapitza Club," an informal gathering of young Cavendish scientists, to discuss the latest discoveries in the field. (His colleagues jested that Kapitza had founded the Club primarily to keep in touch with the advances in physics without having to do the reading himself.[53]) Kapitza remained in England for thirteen years.

Historian Andrew Sinclair has argued that the voluntary and open flow of scientific information from the Cavendish Laboratory proved to the Soviet Union far more vital to the success of the Russian nuclear program than the treason of men such as Fuchs.[54] This is unlikely. When Kapitza went to study with Rutherford, no one believed that nuclear physics had any practical value. Moreover, during his stay in Britain, Kapitza worked primarily on solid state physics rather than nuclear physics.

The issue of correct ideology, however, eventually caught up with Soviet science. During the late 1920s and early 1930s, the Academy lost its intellectual autonomy, and was forced to elect party members to its ranks. When Kapitza travelled to the Soviet Union for a visit in 1934, he was not allowed to return to the West. Upon Stalin's insistence, he headed a new research institute, the Institute for Physical Problems. Kapitza's personal role in the Soviet weapons program, however, has never been made clear. Other Soviet scientists, G.N. Flerov, K.A. Petrzhak and Igor Vasilevitch Kurchatov, also engaged in extensive nuclear research during the 1930s.

During this period the West remained generally less aware of the advances in Soviet physics than vice versa. Historian Arnold Kramish has argued that by the end of the depression decade, the Soviet scientists were equal to their western counterparts, both in theory and in knowledge of experimental techniques.[55]

In early 1942, physicist Georgii Flerov, then a young officer in the Soviet Army, discovered that all the major names in western nuclear physics had ceased writing articles for the scholarly journals in the field. From this he concluded (correctly) that American nuclear research had become secret and was primarily directed toward the building of a bomb. Alerted by "the dogs that do not bark," he wrote Stalin that "we must build the uranium bomb without delay."[56] That same year Stalin appointed Kurchatov as director of the "uranium problem." As historian David Holloway has noted, Stalin ordered the scientists to pursue an atomic bomb in 1942 chiefly because the Soviets knew that the Germans and British and Americans were working on it. The German invasion caused a temporary halt as it forced the Soviets to remove their program to Kazan and other industrial cities beyond the Urals. The dropping of the bombs on Hiroshima and Nagasaki announced the success of the Anglo-American program to the world.

After the end of the conflict, Stalin assigned the uranium project to the KGB. He urged those responsible for uranium research to break the Anglo-American monopoly as quickly as possible.[57] "A single demand of you, comrades," Stalin told his scientists in mid-August, 1945. "Provide us with atomic weapons in the shortest possible time."[58] Interestingly enough, postwar Soviet physics largely escaped the demand for ideological purity that so hampered the growth of Soviet biology. In 1949, Laventy D. Beria, head of the KGB, asked Kurchatov if quantum theory and relativity were not idealist configurations that the Soviets should avoid. Kurchatov replied: "We are working on the A-bomb project now. The A-bomb is based on the theory of relativity and quantum mechanics. Should we discard them, the project would have to be discarded as well." Beria made no further inquiry.[59]

The expertise of the Soviet physicists, the material passed by Klaus Fuchs, and still-to-be-defined assistance from captured German scientists all played a role in this enterprise.[60] On August 29, 1949, the Soviet Union detonated its first weapon. Russia beat Britain to become the world's second nuclear power.

As J. Robert Oppenheimer observed in 1954, of all the known leakages of information, Fuchs's data were, by far, the most significant.[61] The initial Soviet test closely resembled the Trinity device, and the main Soviet isotope separation plant near Sverdlovsk was constructed in the identical U-shape of its Oak Ridge counterpart.[62]

Although western scientists realized that the Soviets would eventually develop a bomb, there is no question that Klaus Fuchs played a considerable role in the timing of the first Soviet weapon. But his exact impact is still open to question. On one side of the issue, reporter Bob Considine suggested that Fuchs saved the Russians ten years; but that is a minority view.[63] Fuchs himself estimated he helped them by "several years." Some observers, however, have argued that the material Fuchs passed was, in the long run, not all that important. Historian H. Montgomery Hyde suggested that Fuchs saved the Soviets approximately one year.[64] Physicist Richard Knightley reduced it to "months." In 1951, *Bulletin of the Atomic Scientists* editor Eugene Rabinowitch observed that it was possible to keep many things secret for a short time and a few things secret for a long time, but not many things secret for a long time. Fuchs, as another historian noted, "merely hastened the end of a process as futile as trying to keep secret the discovery of the wheel."[65]

A consensus on the issue, however, seems to be emerging. In 1950, the Joint Committee on Atomic Energy estimated that Fuchs shortened the Soviet weapons program by at least eighteen months. Sir Rudolf Peierls once asked a group of Soviet scientists how much Fuchs had aided them, and after reflection, they estimated perhaps two years. Historian David Holloway, who interviewed the Soviet scientists who had worked with Fuchs, came up with a similar estimate. Fuchs's 1988 obituary in the *New York Times* also estimated one to two years.[66]

Neither the British nor the American reading public seems to be able to get enough on the infamous Cambridge spies. Books, articles, and plays on Harold "Kim" Philby, Donald Maclean, Guy Burgess, and Anthony Blunt continue to pour from the presses at a steady rate.[67] Seven books on these men appeared in a six-month period during 1988–9. The hunt for "Stalin's Englishmen" has grown into a major indoor British sport.

The story of Klaus Fuchs pales by comparison. A German refugee who arrived penniless in Bristol in 1933 and rose to success because of his scientific gifts is a bourgeois tale. It lacks the mystery, intrigue (and the sexuality) of the Cambridge spies' "betrayal of one's social class."

It does not diminish the treachery of the four Cambridge spies – Philby surely sent several agents to their deaths – to suggest that Klaus Fuchs passed on more significant material than all of them put together. Quite simply, Fuchs changed the flow of history. Because of the data he supplied, the Soviet Union detonated its first atomic explosion in August, 1949, rather than 1951 or 1953. Fuchs's espionage determined the configurations of the early nuclear age. It was all a matter of timing.

After serving nine years of his sentence, Fuchs was set free in 1959. During his term in jail, British officials revoked his citizenship, and he became, literally, a man without a country. Since there was no nation to which he could be deported Fuchs might have remained in Britain for the rest of his days. But his scientific work was his life, and he surely would have been forever barred from any employment worthy of his intellect. For these and for renewed ideological reasons, he decided to fly to East Germany. On his way to Heathrow Airport, he told reporters that he bore neither Britain nor the West any grudge for their actions.

The East Germans treated him well. He was given a position at their chief nuclear facility and in 1974 became its head (ironically, because his superior fled to the West). His adopted country also showered him with honors, including the title of "exceptional scientist of the people" (1980), the East German Order of Merit of the Fatherland (1981) and the country's highest civilian decoration, the Karl Marx Medal. He also was a member of the East German Academy of Sciences. Throughout his new career, Fuchs never resumed contact with any of his former British Mission colleagues, not even the Peierls.

Fuchs also showed no remorse. This may be seen in his initial appearance on East German television in 1960. During the discussion, his colleague, Professor Max Steenbech, declared that the American monopoly on the atomic bomb "had to be broken." "Yes, that had to be broken," Fuchs repeated emphatically after him. That same year he confessed that he "would do it again."[68]

While Fuchs generally refrained from public pronouncements, he did occasionally speak out on world events. He never varied from the official party line. In 1977, he termed the American neutron bomb "an abominable example in the misuse of science." He also attacked the American "Star Wars" Strategic Defense Program. In 1983, he accused American scientists of trying to impose their will on the world through space weapons. In one of his last interviews, he said that Communism was the better system for scientists because it allowed them to translate their sense of responsibility into practice.[69]

On January 28, 1988, Emil Julius Klaus Fuchs died in East Germany. He was 76 years old. Seven months later, at the Nevada Test Site, about sixty Soviet Union and United States officials shared front row seats for the first joint USA/USSR atomic test, a part of the Joint Verification Experiment code-named "Kearsage" for a Sierra Nevada ghost town. After the test proved flawless, spontaneous applause spread through the control room, and US and Soviet officials exchanged handshakes and broad smiles. A few months later, the Americans visited the Soviet test site at Semipalatinsk for a repeat performance.[70] Such are the vagaries of atomic history. One wonders what Klaus Fuchs would have said on this matter.

7 The British Mission and the Postwar Nuclear Culture

Because it could never be put in absolute terms, the impact of the British Mission to Los Alamos has been hard to characterize. Most historians who have dealt with the theme have fallen back on generalities. Stephane Groueff, for example, termed the British/American Los Alamos collaboration "perfect," and the overall British Mission contribution "inestimable."[1] Andrew Pierre suggested that their main contribution lay in "uncommensurables," such as providing secondary opinions, special equipment, and special talents that the Americans lacked.[2] Hewlett and Anderson have noted that the Mission "lent an international atmosphere" to Los Alamos.[3] Peter Malone has confessed that their contribution was "difficult to assess."[4]

The British scientists themselves have remained equally vague in describing their role in the creation of the atomic bomb. Sir Michael Perrin acknowledged only "significant contributions" at Los Alamos while Sir John Cockcroft called the overall British effort "an effective though belated contribution" to the US bomb.[5] A.P. French suggested that the handpicked group, with its core of genuinely distinguished people, probably served as "a little leavening, as it were, here and there, of the much bigger effort being done by the Americans."[6] P.B. Moon felt the major British contribution lay with the Frisch-Peierls Memorandum and the Maud Report. He considered their greatest technical contribution at Los Alamos to be the prediction of the mechanical effects of the explosions by Taylor, Tuck and Penney.[7] Sir Rudolf Peierls considered the question of the overall impact of the British Mission too complex, and declined to estimate their impact.[8]

Those who worked directly with the Los Alamos British team, however, were lavish with their praise. Los Alamos veterans Ralph Carlisle Smith and Robert Porton recalled the British Mission as all superior people. Norris Bradbury termed them "an extremely able group, every one." He especially lauded Penney for his contributions. After the war, Oppenheimer also went out of the way to credit their accomplishments.[9]

Even General Groves gave way grudgingly. In 1954, he told the Atomic Energy Commission that the British were "a scientific reservoir." In 1949, he wrote privately that "the main British contribution was in arousing and maintaining the interest and enthusiasm of President Roosevelt in the project. This was of real value. Among other things, it was probably the major factor in our keeping top priority throughout the war."[10]

In 1949, Hans A. Bethe wrote to Carroll L. Wilson, General Manager of the AEC, to note that the British contributions to the Theoretical Division during World War II were "absolutely essential." "It is very difficult to say what would have happened under different conditions," Bethe continued. "However, at least the work of the Theoretical Division would have been very much more difficult and very much less effective without the members of the British Mission, and it is not unlikely that our final weapon would have been considerably less efficient in this case."[11] Official recognition came in 1946 when the American government awarded Peierls, Penney, and Frisch the American Medal of Freedom for their work at Los Alamos.

Lord Penney has suggested that any estimate of what would have happened "if the British scientists had not been at Los Alamos" must, of course, be speculative. But in 1987, John Manley ventured to put a precise number on the British contribution. Manley estimated that the group was 20–40 per cent more effective than any random group could have been, and 50 per cent more effective than any similar American group.[12] Upon further reflection, A.P. French observed that the contributions made by Frisch, Peierls, Penney, Taylor, Titterton, and Tuck were all out of proportion to their numbers.[13] Penney noted that while there were American scientists who could have replaced most of the British on a man-for-man basis, they were scattered in various laboratories and it is unlikely that Los Alamos could have recruited them all. Penney stated that the British scientists at Los Alamos made "an important but small contribution" to the weapons work there. His intuitive assessment was around ten per cent.[14]

The Frisch-Peierls Memorandum, the Maud Report, Churchill's constant pressure on Roosevelt, and the Los Alamos "ten per cent" all accelerated the ultimate goal: production of a weapon "in the shortest possible time." Thus, the British Mission to Los Alamos was essential in determining the precise date of the world's first atomic bombs. One American scientist believed the British contribution shortened the project by about a year. Without the British, Margaret Gowing has argued, World War II would have surely ended before a bomb could have been utilized. For better or worse, atomic weapons would not have been ready in August 1945 without them.[15]

But the story does not stop with the tragedy of Hiroshima and Nagasaki. Members of the British Mission to Los Alamos also helped determine the timing of the first Russian test in August, 1949, and the first British test in October, 1952. The first three nuclear powers all had direct links to the

British Mission to Los Alamos. Collectively, they changed the flow of history.

The impact of their Los Alamos experience, moreover, spread all through the postwar era. The creation of an independent British nuclear deterrent, and Fuchs's aid to the Soviet program have already been alluded to. But one should also mention three additional postwar items: the impact of the Pugwash Conferences; the public fascination with nuclear espionage; and the first generation of scientists' assessment of their own creation.

At Los Alamos, Niels Bohr had argued vigorously that nuclear matters belonged in an "open world" of international science. Scientists had long assumed that scholarship was "international." For several years after the war, the theme that "science can flourish only in an atmosphere of free interchange of ideas" also dominated the American scientific community. Virtually every scientist of note spoke out on the need for free international exchange of scientific data.[16] The strange course of postwar politics, however, made that increasingly difficult.

But Bohr's idea concerning openness in scientific matters did find a home in Pugwash. The Pugwash Conferences were established in 1955, largely through the efforts of mathematician and philosopher Bertrand Russell, Albert Einstein, and the British Atomic Scientists Association. Pugwash emerged as the scientists' response to the escalating arms race between the US and the Soviet Union, especially over the dangers of radioactive fall-out. The founders hoped that the Pugwash gatherings would provide the best chance to continue the international openness that had characterized science in the pre-World War II years in Europe. In the 1955 Pugwash Magna Carta, written by Russell, and co-signed by ten others, the scientists said: "In the tragic situation which confronts humanity, we feel that scientists should assemble in conference to appraise the perils that have arisen as a result of the development of weapons of mass destruction." The manifesto concluded:

There lies before us, if we choose, continued progress in happiness, knowledge and wisdom. Shall we, instead, choose death because we cannot forget our quarrels? We appeal, as human beings, to human beings. Remember your humanity and forget the rest. If you can do so, the way lies open to a new Paradise; if you cannot, there lies before you the risk of universal death.[17]

The organizers soon found a sympathetic ear in Cyrus Eaton, a Canadian-born American industrialist. In 1955, Eaton had opened his home in Pugwash, Nova Scotia, to scientists from twenty-two nations and assumed most of the participants' expenses (hence the name). This was the beginning. At their gathering in 1957, the scientists issued their famous "Vienna Declaration." This stated that scientists across the world should strive to educate both public opinion and political leadership on the vital issues of the atomic age. This has been a central focus of the organization up to the present. It has been endorsed by thousands of scientists around the world.

The Pugwash gatherings involved hand-selected scientists from all over the globe. They usually met in closed sessions and the twenty to forty people involved soon got to know each other well. The scientists agreed to discuss, on a no-holds-barred basis, all the crucial issues of the day: the arms race, disarmament, arms control, world security, chemical weapons, and numerous other international issues. Since the 1957 gathering called for immediate suspension of nuclear testing, neither the American nor the Soviet government ever encouraged the organization.

The general public has never been well acquainted with the Pugwash conferences. The closed sessions have prevented widespread publicity, and the scientists involved have always been ambivalent about public exposure. But the conferences have created an environment and atmosphere in which scientists from both East and West could speak out freely on all the issues of the day. Arms control, chemical warfare, disarmanent – nothing was off limits. The scientists hammered out the problems and then reported back to the decisionmakers of their own governments.

Although Pugwash conferences met outside the glare of publicity, their role may be larger than has been generally realized. Pugwash had special impact on the postwar leaders of the Soviet Union. In the early days of the arms race, the Soviets believed that if a nuclear conflict erupted between East and West, the East would emerge as eventual winners. There might be suffering, to be sure, but history dictated the eventual victory of the proletariat. Even Nikita Khrushchev held to this position. Thus, the Pugwash group of Soviet scientists helped educate the Kremlin to the now-accepted truth that neither side could hope to win a nuclear war.[18]

If the original idea for Pugwash came from Niels Bohr, the central figure in its operation has been Joseph Rotblat. Not only was he one of the original organizers, Rotblat also served as a bulwark of Pugwash for over three decades. Doggedly he kept the faith, even when East–West relationships were at their most strained: the Berlin Wall, the Cuban Missile Crisis, Vietnam, and Afghanistan. For seventeen years he was General Secretary and he also wrote the only history of the movement.[19]

With the rise of *perestroika* in the late 1980s, both Rotblat's and Pugwash's stars have steadily risen. In 1989, Roblat was feted as the only person to have attended all the Pugwash conferences since their founding.

On November 4, 1988, the British Pugwash group held a special London symposium "The Future of Pugwash" in honor of Rotblat's eightieth birthday. As "a founding Pugwashite," Joseph Rotblat had become the symbol of the movement. As he said in a recent interview, "It is because we approach political problems in the objective spirit of science that governments have come to listen to us and trust us."[20]

While the rise of international terrorism has placed a permanent hold on Bohr's hope of openly sharing nuclear data, his dreams of exchanging scientific information – the "open world" that his Copenhagen Institute so embodied – came at least to partial fruition with Pugwash.

A second aspect of the postwar world with a distinct British Mission link may be seen in the American national obsession with "security" during the 1950s and 1960s. Immediately after the war, both British and American scientists devoted much time to dispelling the popular concept that there was a simple "secret," or atomic "formula" to the bomb. A week after Nagasaki, Sir James Chadwick said that any industrialized nation could produce an atomic bomb within five years, without any further assistance. It would largely be done, Chadwick observed, "just knowing that it is possible:"[21] A few months later, Hans Bethe warned a Denver audience that the Russians would have a nuclear weapon within three to six years, regardless of whether they were given any information or not.[22] In *One World or None* (1946), Bethe laid out the step-by-step processes by which the Soviets could do this.[23] As Marcus E. Oliphant told a Rotary Club audience in September, 1945, the idea that any country could keep the nuclear process secret was "just rot."[24] The only secret of the atomic bomb, Niels Bohr was quoted as saying, "is that it can be built."[25]

This scientific truism, however, failed to dent the American public mind. Mass politics moved on a far less rational and much more complicated plane than did scientific experiments. For years the public remained convinced that there was a "secret," that it could be "stolen," and that America was foolish to "share" it voluntarily.

This seemed like common sense to most Americans. Shortly after the end of the war, the former governor of Michigan wrote President Truman urging that Truman save the world by "burying" the "secret" of the atomic bomb on an island near Sault Ste Marie. UN soldiers could then guard the "crypt" twenty-four hours a day.[26] In April, 1946, a Washington attorney killed himself and shot his family because he feared that America's atomic bomb secrets were being leaked to an "enemy."[27] In 1947, when six GIs stationed at Los Alamos took bomb casing photographs home as souvenirs, they were arrested and accused of betraying atomic secrets. After this,

General Groves tried to reassure the nation that no spy could "steal" the bomb. "The bomb project is like a giant jigsaw puzzle with 100,000 pieces," Groves said. "A spy ring might possibly get five or ten pieces out of 100,000 but it wouldn't do it much good."[28]

Few seemed convinced. The rise of science fiction during the 1930s had introduced the concept of a "secret formula" to the American public, and this idea did not easily fade. Cartoonists and journalists in the 1950s used exactly the same phrase.[29] After Klaus Fuchs's espionage became known – without any specifics, of course – Fuchs became a quasi-mythical figure: the "atomic spy" who gave the "secret" of the bomb to the Soviet Union (perhaps even to the Chinese). American obsession with nuclear espionage in the postwar decades traces back, in part, to the British Mission to Los Alamos.

A corollary to the popular suspicion may be seen in the American security forces' continual distrust of their British counterparts, MI5 and MI6. After the exposure of Fuchs, security officials in both nations engaged in serious discussion whether proper vetting would have uncovered his early communist connections. This, of course, is still a matter of conjecture. In the mid-1930s, the German Consul in Bristol had told the local chief constable that Fuchs was a communist. That information was passed on to British Security, but it was dismissed as "tainted evidence." As Prime Minister Clement Attlee noted in Parliament, all opponents of the Nazis were termed "communists." MI5 argued that Fuchs had shown no further involvement with the party while in Britain. Confiscated Gestapo records from Kiel revealed his party involvement, even his membership number, but those data were somehow lost. In all, Fuchs was cleared by British intelligence a total of eight different times.

Hindsight is always 20/20, but numerous recent accounts have suggested that the British were indeed careless in their examination of the Fuchs case. Although there were exceptions, most non-Jewish refugees from Germany were quite left wing. A recent reminiscence recalled that Fuchs was well known as an active member of the Bristol communist cell.[30] Max Born noted that when Fuchs arrived in Edinburgh, he was a dedicated communist. Born allegedly told Fuchs not to engage in such activities in Scotland, and Fuchs then became his shy, retiring self.[31] In 1948, an MI5 officer reviewed the Fuchs materials and concluded that Fuchs had pro-bably been a spy.[32] Yet nothing was done until the FBI alerted MI5 about the recently cracked KGB messages.

Groves accepted the MI5 clearance of British Mission members. To have insisted on rechecking everyone would have insulted a wartime ally. In fact, J. Edgar Hoover told a Congressional committee in 1950 that if MI5 were to clear a similar group today, identical security arrangements would apply.[33]

Because Fuchs sailed through these security checks so easily, some observers have concluded that he must have had top-level assistance. Former MI5 officer Peter Wright has argued that Sir Roger Hollis, the Head of MI5, was the "fifth man" who aided Fuchs. Historian Nigel West has placed the suspicion on Graham Mitchell, a deputy of Hollis. In September, 1987, Yuri Modin, a KGB official, confessed that the Soviets did, indeed, have a "fifth man" in British intelligence circles. When asked by a British reporter to identify him, Modin said only: "British counter-intelligence has been trying to work that one out without success for nearly 30 years now."[34]

Since both Hollis and Mitchell are dead, the issue is unlikely to be soon resolved. But the names of Fuchs, Philby, Burgess, Blunt, and Maclean still bring shudders to American intelligence officers. That, too, is a legacy of the British Mission to Los Alamos.

Finally, the men and women from the British Mission to Los Alamos have wrestled for almost half a century with the implications of their discoveries. When a reporter once asked Hans Bethe about the impact of Los Alamos, he replied: "It changed everything; it took scientists into politics." After August of 1945, the average citizen realized the implications of atomic weapons almost immediately.[35] Scientists had assumed a new responsibility on the world's stage.

As a prominent American psychologist observed, the splitting of the atom has impinged on all structures and functions of the modern social order.[36] No aspect of modern life remains untouched. The post-World War II generation has used every creative means at its disposal to grapple with the issues raised: politics, photography, popular music, theatre, films, novels, poetry, sculpture, theology, history, science, political philosophy and so on.

The main arena for discussion, of course, has been the political one. Although every issue of the *Bulletin of the Atomic Scientists* methodically lists the American or Soviet weapons stockpiles (over 20,000 each), the nuclear powers have made at least some progress in bringing the arms race under control. The 1963 Partial Test Ban has kept the superpowers from above-ground testing, even if it hasn't halted weapons development; the Salt I and Salt II treaties have brought about slight reductions on certain types of weaponry and the ABM treaty has prevented deployment of antiballistic missile systems.[37] Agreements not to militarize space or Antarctica are also hopeful signs. A 1990 Department of Defense proposal announced plans to reduce nuclear weapons to a minimum of 3,000 by the twenty-first century.[38] The advent of *perestroika* has raised cautious hopes everywhere.

In *Danger and Survival* (1985) historian and politician McGeorge Bundy has argued that the most important legacy of the Tube Alloys/Manhattan Project has been the continuing tradition of *non-use* of nuclear weapons. While developing the bomb has been viewed as necessary for "great power" status, the practical, day-to-day utility of the weapons has been very small.[39] Since regression to a virginal, pre-1945 state is not possible, all future politicians will have to confront this dilemma with each new generation.

In addition to the political aspect, ecologists and psychologists have also recently made their voices heard. During the 1980s, the concepts of "nuclear winter" and acute medical and "psychological dislocation" in the face of a nuclear attack were both much in the news.

In April 1983, about one hundred scientists met in Cambridge, Massachusetts, to discuss the concept of nuclear winter and the results were publicly announced on October 31 and November 1, 1983. *Science* published "Global Atmosphere Consequences of Nuclear War" in its December 23, 1983 issue.[40] In the ensuing months, scholarly publications vied with popular presentations, such as Jonathan Schell's *The Fate of the Earth*, to bring this idea to the public. Astronomer Carl Sagan has emerged as the chief spokesman of this position, with lucid articles in such wide-ranging journals as *Foreign Affairs* and *Parade* magazine. British, American, and Soviet scientists all agree on the dangers of nuclear winter. The idea is not a cold war item. With the discovery that nuclear war could trigger global ecological catastrophe, Sagan has argued, pre-1945 wisdom cannot be relied on as an adequate guide to sane political behavior.[41] The call is for new, creative thinking on all fronts.

During the same period, the medical profession responded with equal vigor. In 1982, a gathering of the nation's physicians informed President Ronald Reagan that there could be no adequate medical care in the wake of even a small nuclear exchange. Physicians for Social Responsibility, aided by the writings of Australian physician, Helen Caldicott, have echoed this position.[42]

The psychologists and psychiatrists have also contributed to the discussion. In the early 1980s, both British and American psychological associations held conferences on the psychological aspects of nuclear conflict. A British researcher has argued that the most probable psychological results of a nuclear war would not be panic but "disaster victim syndrome," i.e. apathetic, docile, indecisive and mechanical responses. American psychiatrist Rita Rogers has suggested that the world is unable to provide a fear commensurate with the nature of the threat at hand. Her colleague has argued for a renewed "courage to be afraid."[43]

American psychiatrist Robert Jay Lifton has emerged as the chief spokesman for this aspect of nuclear culture. In *Death in Life* (1968), *Indefensible Weapons* (1982) and elsewhere, Lifton has maintained that the rising scale of nuclear arms build-up has psychologically impaired all human thought. It has cut humankind off from its most precious asset, he argues, for it has impaired our imagination of the future.[44] The world possesses no symbol adequate to the proposed danger of the extinction of humanity. Those that come closest, perhaps, are the religious images of Armageddon drawn from the books of Daniel and Revelation. But, they always envisioned the end of history as a redemptive event, an act of the divine for His own purposes. Yet a man-made nuclear holocaust involves death in and of itself; there can be no redemptive purpose. As Lifton put it: "When the image of nuclear winter threatens our symbolization of immortality, then, it threatens a level of psychic experience that defines our humanity."[45] Across the Atlantic, the British Campaign for Nuclear Disarmament (CND) has echoed similar themes. Their Easter, 1958, parade drew 10,000 people in the longest procession ever seen on English roads.

The men and women who were at Los Alamos from 1943 to 1945 have accepted a degree of responsibility for this situation. Their joy at the ending of the war was always tempered by the sadness of Hiroshima and Nagasaki. Perhaps this "first generation" of atomic scientists, the group that actually built the bombs with their own hands and witnessed the Trinity test in person, had a deeper understanding of their creation than their successors. As weapons production became standardized and younger people took the helm, the political decisionmakers moved further from the common assumptions that the "first team" had all shared. With few exceptions, the British members of this "first team" carried home a decided sense of ambivalence from their Los Alamos days. D.J. Littler, for example, declined to become involved with the production of Britain's atomic weapons. Instead, he devoted his career to the problems of energy production. While James Tuck originally returned to Los Alamos to assist Edward Teller with the development of the hydrogen bomb, he soon broke with Teller and devoted the brunt of his later career to work on the Sherwood Project, an attempt to provide commercial power from fusion. Tuck also confessed that his only recurring dream involved fleeing from an encroaching mushroom cloud. Elsie Tuck remained convinced that all the people at Los Alamos experienced some sense of guilt for launching the bomb on the world.[46]

This ambiguity, perhaps, is more clearly reflected in the Continental members of the British Mission than in those who were Island born. Frisch avoided politics completely. In private, Bohr frequently regretted the use of the bombs. Rotblat did so publicly. The first release of nuclear energy occurred in a reactor in Chicago, Rotblat observed. But the world learned about nuclear power only through the announcement of Hiroshima. This linkage, he argued, can never be erased. Atomic energy came to birth with a stigma attached to it.[47]

During the Los Alamos days, however, the Hill bristled with discussions about how the new weapons would be so horrendous they "would bring the world to its senses." The arguments for increased stockpiles have echoed the same themes. Nuclear weapons were made and tested, Lord Penney told Lorna Arnold, in the solemn hope that they would make world war impossible in the future.[48] A number of his Los Alamos colleagues shared that assumption. Hans Bethe argued that nuclear weapons saved the peace in Europe immediately after the war. J. Robert Oppenheimer also credited them for the uneasy international stability of the mid-1960s. Long-term director of Los Alamos, Norris Bradbury, has evolved a stock reply to this question: "We've had no world war for fifty years; we must have done something right."[49]

And yet the doubt remains. "There hasn't been a [nuclear] war yet," Oppenheimer told a *Newsweek* reporter in 1965, "I say with a great deal of knocking on wood." But over a decade later the rapid proliferation of weapons called several Los Alamos veterans out of retirement. Former British Mission member J. Carson Mark has half seriously suggested that all heads of state be forced to witness one nuclear explosion each year. Hans Bethe felt the same way. Recently Bethe observed that his generation of scientists assumed that postwar national arsenals would contain only a few dozen atomic weapons each – if they had any at all.[50] No one dreamed of national arsenals with over 20,000 nuclear weapons.

In Britain, Joseph Rotblat and Sir Rudolf Peierls have both echoed these themes. Long-term veterans of the British Atomic Scientists' Organization, in the early 1980s they became active in the Nuclear Freeze movement. As part of their witness, they joined other scientists to stage a vigil on the steps of St Martin-in-the-Fields in Trafalgar Square in protest against "the misuse of nuclear weapons." As Sir Rudolf Peierls observed in 1986, "I'm afraid, forty years ago, we over-estimated the capacity of those in power to understand the implications of what we had created."[51]

Notes

The following abbreviations have been used throughout the notes below (see also the primary sources listed in the Bibliography at the end of the book):

FDR Library   Franklin D. Roosevelt Library, Hyde Park, New York
FRUS   *Foreign Relations of the United States* (Washington: USGPO, 1968)
HST Library   Harry S Truman Library, Independence, Missouri
KF   Klaus Fuchs papers
LANL Archives   Los Alamos National Archives, Los Alamos, New Mexico
PRO   Public Record Office, Kew, London

INTRODUCTION

1. H.G. Nicholas (ed.), *Washington Dispatches, 1941–45* (Chicago: University of Chicago Press, 1981), pp. 598–9.
2. John Franklin Carter to Matthew J. Connelly, August 7, 1945, Papers of Harry S Truman, Official File Box 1527 Misc (1945), HST Library.
3. Henry DeWolf Smyth, *Atomic Energy for Military Purposes: The Official Report on the Development of the Atomic Bomb Under the Auspices of the United States Government, 1940–1945* (Princeton: Princeton University Press, 1945); H.D. Smyth, 'The "Smyth Report,"' *The Princeton University Library Chronicle*, Vol. XXXVII (Spring, 1976), pp. 173–89; cf. Waldemar Kaempffert, 'Story of Scientists; "Battle" for Atom Bomb Secrets Revealed in Smyth Report,' *New York Times*, August 16, 1945, p. 8.
4. *Statements Relating to the Atomic Bomb* (London: His Majesty's Stationery Office, 1945). While the White Paper is out of print, an abbreviated 'Statements by the Prime Minister and Mr. Churchill on the Atomic Bomb, 6 August 1945' may be found in Margaret Gowing, *Independence and Deterrence: Britain and Atomic Energy, 1945–1953*, Vol. I: *Policy Making* (London: Macmillan, 1974), pp. 14–18.
5. McMillan to Longair, October 18, 1945, ABI/54, PRO.
6. N. Feather, 'American Work on the Atomic Bomb Project,' *Nature*, Vol. 156 (December 29, 1947), pp. 768–9. 'The Publishing History of the "Smyth Report,"' *Princeton University Library Chronicle*, Vol. XXXVII (Spring, 1976), pp. 191–201.
7. W.A. Akers to J.P. Baxter, October 11, 1945. AB1/54, PRO.
8. 'The Beginning or the End,' *Nature*, Vol. 159 (April 26, 1947), p. 583.
9. Ronald W. Clark, *The Birth of the Bomb: The Untold Story of Britain's Part in the Weapon That Changed the World* (London: Phoenix House, 1961); Clark admitted, however, that the British contribution could not be easily

summarized. While he suggested that it was not accurate to claim that the British 'thought' the bomb and the Americans 'built' it, his book argues along precisely those lines (p. xxiii).

10. Margaret Gowing, *Britain and Atomic Energy, 1939–1945* (New York: Macmillan, 1964); Margaret Gowing to author, November 19, 1985.

11. James Phinney Baxter III, *Scientists Against Time* (Cambridge, Mass.: MIT Press, 1968) [1946]), pp. 419–47.

12. Leslie R. Groves, *Now It Can Be Told: The Story of the Manhattan Project* (New York: Harper, 1962; 1983), p. 408. Groves to Strauss, October 13, 1949. Strauss papers, Herbert Hoover Presidential Library, West Branch, Iowa. Copy lent by Barton J. Bernstein.

13. *Manhattan Project: Official History and Documents*, Book VIII, Los Alamos Project, Part 16, 'Diplomatic History of the Manhattan Project,' National Archives Microfilm (1977) Reel 10, pp. 42–5. Parts of this study were published in Anthony Cave Brown and Charles B. MacDonald (eds), *The Secret History of the Atomic Bomb* (New York: Dial Press, 1977).

14. R.V. Jones, *The Wizard War* (New York: Coward, McCann and Geoghegan, 1978); James Phinney Baxter III, *Scientists Against Time* (Cambridge, Mass.: MIT Press, 1968 [1946]); and David E. Fisher, *A Race on the Edge of Time* (New York: McGraw-Hill, 1988).

15. Margaret Gowing, 'Reflections on Atomic Energy History,' *Bulletin of the Atomic Scientists*, Vol. 35 (March, 1979), p. 51.

16. Interview with Peggy Titterton (telephone), January 14, 1989. Copy in LANL Archives.

17. *New York Times,* July 15, 1970 (Groves obituary); *cf. New York Times,* February 19, 1967 (Oppenheimer obituary).

18. Interview with Peggy Titterton. Rudolf Peierls, felt the same way. See Sir Rudolf Peierls, 'J. Robert Oppenheimer,' *Dictionary of Scientific Biography*, Vol. X (New York: Scribner's, 1984), pp. 213–18.

19. Bradbury, as cited in Ferenc Morton Szasz, *The Day the Sun Rose Twice: The Story of the Trinity Site Nuclear Explosion, July 16, 1945* (Albuquerque: University of New Mexico Press, 1984), p. 19. The best studies of Oppenheimer are Peter Goodchild, *J. Robert Oppenheimer: Shatterer of Worlds* (Boston: Houghton Mifflin, 1981); and James W. Kunetka, *Oppenheimer: The Years of Risk.* See also Alice Kimball Smith and Charles Weiner (eds), *Robert Oppenheimer: Letters and Recollections* (Cambridge, Mass.: Harvard University Press, 1980). The hearings (with a valuable index) are available: *In the Matter of J. Robert Oppenheimer* (Cambridge, Mass.: MIT Press, 1979). Tufts University Professor Martin Sherwin is currently working on a full-scale biography of Oppenheimer.

20. K.D. Nichols, *The Road to Trinity: A Personal Account of How America's Nuclear Policies Were Made* (New York: William Morrow, 1987), pp. 99–111.

21. M.L. Oliphant, 'Notes on conversations with the Americans in Washington, at the New War Department, on Monday, September 13, 1943,' AB1/376, PRO.

22. Groves, as cited in Barton J. Bernstein, 'The Quest for Security: American Foreign Policy and International Control of Atomic Energy, 1942–1946,' *Journal of American History*, Vol. LX (March, 1974), p. 1004.

23. The two best sources are his autobiography, *Now It Can Be Told*, and William Lawren, *The General and the Bomb: A Biography of General Leslie R. Groves, Director of the Manhattan Project* (New York: Dodd, Mead, 1988).

24. Powell, cited in Necah Stewart Furman, *A History of Sandia National Laboratories – The Postwar Decade* (Albuquerque: University of New Mexico Press, 1990), p. 125. Lawren, *The General and the Bomb*, p. 285.

25. See *The Journals of David E. Lilienthal, The Atomic Energy Years, 1945–1950)*, Vol. II (New York: Harper, 1964) on this.

26. Statement by Truman, Subject file, Memoirs; Foreign Policy/A Bomb, HST Library.

27. William Lawren, *The General and the Bomb, passim.*

28. Margaret Gowing, *Britain and Atomic Energy, 1939–1945*, p. 262n.

29. Although a superb reference librarian, she remains nameless at her own request.

CHAPTER 1   BACKGROUND

1. Otto Hahn, "The Discovery of Fission," *Scientific American*, Vol. 198 (February 1958), pp. 76–84. For the French role in this saga, see Bertrand Goldschmidt, *The Atomic Adventure*, translated from the French by Peter Beer (Oxford: Pergamon, 1964) and Spencer R. Weart, "Scientists in Power: France and the Origins of Nuclear Energy, 1900–1950," *Bulletin of the Atomic Scientists* (March, 1949), pp. 41–50.

2. Lise Meitner, "Right and Wrong Roads to the Discovery of Nuclear Energy," *International Atomic Energy Bulletin* (December 2, 1982), pp. 6–8; Otto R. Frisch, "How It All Began," *Physics Today*, Vol. 20 (November 1967), pp. 43–8; cf. Lise Meitner, "Looking Back," *Bulletin of the Atomic Scientists* (November 1964) and Otto R. Frisch, "Experimental Work with Nuclei: Hamburg, London, Copenhagen," in Roger H. Stuewer (ed.), *Nuclear Physics in Retrospect: Proceedings of a Symposium on the 1930s* (Minneapolis: University of Minnesota Press, 1979), p. 71–3.

3. Lise Meitner and O.R. Frisch, "Letter to the Editor," *Nature*, Vol. 143 (February 11, 1939), pp. 239–40.

4. *The Times*, November 16, 1979.

5. Cited in Roger H. Stuewer, "Bringing the News of Fission to America," *Physics Today*, Vol. 38 (October 1985), p. 51.

6. James Phinney Baxter III, *Scientists Against Time* (Cambridge, Mass.: MIT Press, 1946; 1968), p. 420.

7. Cf. George A. Cowan, "A Natural Fission Reactor," *Scientific American*, Vol. 235 (July, 1976), pp. 39–47.

8. John W. Wheeler-Bennett, *John Anderson, Viscount Waverley* (New York: St. Martin's Press, 1962), quoted on p. 289; Frisch, "How It All Began," p. 52; cf. Edward Teller, *Better a Shield Than a Sword: Perspectives on Defense and Technology* (New York: Macmillan, 1987); Ruth Moore, *Niels Bohr: The Man, His Science, and the World They Changed* (Cambridge, Mass.: MIT Press, 1985), p. 298. The best study of the onset of the war is Donald Cameron Watt, *How War Came: The Immediate Origins of the Second World*

War, 1938–1939 (New York: Pantheon, 1989).

9. Winston S. Churchill, *The Gathering Storm* (Boston: Houghton Mifflin, 1948), pp. 386–7; Ronald W. Clark, *Tizard* (Cambridge, Mass.: MIT Press, 1965), pp. 204–6; 211.

10. Cf. Bernard Wasserstein, *Britain and the Jews of Europe, 1939–1945* (Oxford: Clarendon Press, 1979) and Bernard Wasserstein, "Intellectual Emigrés in Britain, 1933–1939," in Jarrell C. Jackman and Carla M. Borden (eds), *The Muses Flee Hitler: Cultural Transfer and Adaptation, 1930–1945* (Washington: Smithsonian Institution Press, 1983), pp. 248–57. Francis Harry Hinsley, *British Intelligence in the Second World War: Its Influence on Strategy and Operation* (London: HMSO, 1979), Vol. I, pp. 116–26; Clark, *Tizard*, p. 204.

11. Rudolf Peierls review of Richard Rhodes, *The Making of the Atomic Bomb, New York Review of Books* (November 5, 1987), p. 47; Clark, *Tizard*, p. 214.

12. R.W. Reid, *Tongues of Conscience: War and the Scientists' Dilemma* (London: Readers Union Constable, 1970), pp. 141–2.

13. McGeorge Bundy, *Danger and Survival: Choices about the Bomb in the First Fifty Years* (New York: Random House, 1988), p. 25.

14. Margaret Gowing, (London: Macmillan, 1964) p. 42; James Chadwick also independently realized the potential for a fast neutron atomic weapon. Mark Oliphant, "James Chadwick," *Physics Today*, Vol. 27 (October 1974), pp. 87–8; Hans Bethe and George Winter, "Otto Robert Frisch," *Physics Today*, Vol. 33 (January 1980), pp. 99–100; Robert W. Clark, *The Birth of the Bomb: The Untold Story of Britain's Part in the Weapon That Changed the World* (London: Phoenix House, 1961), pp. 50–1; Rudolph Peierls, *Bird of Passage: Recollections of a Physicist* (Princeton: Princeton University Press, 1986), pp. 152–5.

15. Cf. Bertrand Goldschmidt, *The Atomic Adventure*, translated from the French by Peter Beer (Oxford: Pergamon Press, 1964), p. 16n. Goldschmidt mistakenly termed Maud "The Military Application of Uranium Disintegration."

16. Gowing, cited in John Baylis, *Anglo-American Defense Relations, 1939–1984: The Special Relationship* (New York: St. Martin's Press, 1984), p. 16; the original Committee members were Professors George Thomson (Chair), James Chadwick, John Cockcroft, Mark Oliphant, and Dr P.B. Moon. During the war both Moon and Chadwick went to Los Alamos, Cockcroft to Montreal, and Oliphant to Berkeley.

17. Peierls, quoted in Clark, *Birth of the Bomb*, 53; FDR to Churchill, October 11, 1941, *FRUS*, p. 3.

18. John Simpson, *The Independent Nuclear State: The United States, Britain and the Military Atom* (New York: St. Martin's Press, 1983), quoted on p. 21.

19. A.J.R. Groom, *British Thinking About Nuclear Weapons* (London: Francis Pinter, 1974), p. 3.

20. Margaret Gowing, "Nuclear Weapons and the 'Special Relationship,'" in William Roger Louis and Hedley Bull (eds), *The 'Special Relationship': Anglo–American Relations Since 1945* (New York: Oxford, 1986), pp. 118–19; cf. J.R. Oppenheimer, "Niels Bohr and Atomic Weapons," typescript, HST Library.

21. Winston S. Churchill, *The Grand Alliance* (Boston: Houghton Mifflin, 1950), pp. 814–15; Wheeler-Bennett, *John Anderson, Viscount Waverley*, p. 290. cf.

Norman Moss, *The Politics of Uranium* (New York: Universe Books, 1982), p. 11.
22. Gerald Pawle, *The War and Colonel Warden* (London: Harrap, 1963), p. 138.
23. Groves, Memorandum, 23, April 1945, as illustrated in Roger M. Anders, "The President and the Atomic Bomb: Who Approved the Trinity Nuclear Test?" *Prologue*, Vol. 20 (Winter 1988), p. 280; K.D. Nichols, *The Road to Trinity: A Personal Account of How America's Nuclear Policies Were Made* (New York: William Morrow, 1987), p. 40. Warren F. Kimball (ed.), *Churchill and Roosevelt: The Complete Correspondence*, Vol. I (Princeton: Princeton University Press, 1984), p. 249.
24. David Brinkley, *Washington Goes to War* (New York: Alfred A. Knopf, 1988), p. 50. See also, Alex Danchev, *Very Special Relationship: Field–Marshall Sir John Dill and the Anglo-American Alliance 1941–44* (London: Brassey's Defence Publishers, 1986).
25. Baylis, *Anglo-American Defense Relations, 1939–1984*, pp. 5–6; David E. Fisher, *A Race on the Edge of Time: Radar – The Decisive Weapon of World War II* (New York: McGraw-Hill, 1988), pp. 268–9; Jack Niessen, with A.W. Cockerill, *Winning the Radar War, 1939–1945* (New York: St. Martin's Press, 1987), pp. 94–101.
26. Steward Cockburn and David Ellyard, *Oliphant: The Life and Times of Sir Mark Oliphant* (Adelaide: Axiom Books, 1981), p. 89.
27. Glenn Fowler, as cited in Necah Stewart Furman, *A History of Sandia Laboratories – The Postwar Decade*, Albuquerque: University of New Mexico Press, 1990), p. 147.
28. John T. Connor to Carroll L. Wilson, February 3, 1943 summarized the foreign liaison activities of the NDRC and OSRD. Copy in Atomic Bomb Folder, FDR Library. Vincent C. Jones, *Manhattan: The Army and the Atomic Bomb* (Washington: USGPO, 1985), pp. 227–8; Peierls, *Bird of Passage*, p. 169; *Manhattan Project: Official History and Documents*, Book VIII, Part 16, "Diplomatic History of the Manhattan Project," National Archives Microfilm, Reel 10.
29. Peierls, *Bird of Passage*, p. 169; Chadwick quoted in Gowing I, p. 85.
30. John S. Rigden, *Rabi* (New York: Basic Books, 1987), p. 148.
31. Cockburn and Ellyard, *Oliphant*, pp. 104–5.
32. Spencer R. Weart and Gertrude Weiss Szilard (eds), *Leo Szilard: His Version of the Facts*, Cambridge, Mass.: MIT Press, 1980), p. 146. In fact, in 1946, General Groves put forth Oliphant's name for the Congressional Medal of Freedom with Gold Palm, the highest honor the States could bestow on a foreigner. However, the Australian government did not permit foreign decorations for Australians, and Oliphant never received the citation. He remained unaware that the honor had been proposed until 1980. Lorna Arnold, *A Very Special Relationship: British Atomic Weapon Trials in Australia* (London: HMSO, 1987), pp. 28–9.
33. Peierls, *Bird of Passage*, pp. 170–1.
34. Danchev, *Very Special Relationship*, p. 101.
35. Barton J. Bernstein, "The Uneasy Alliance: Roosevelt, Churchill, and the Atomic Bomb, 1940–1945," *Western Political Quarterly*, Vol. 29 (1976), pp. 204–5.
36. Bernstein, "The Uneasy Alliance," p. 207.

37. Gowing, "Nuclear Weapons and the Special Relationship," p. 119.
38. Cited in Bernstein, "The Uneasy Alliance," p. 209.
39. Extract of letter from Akers to Perrin, January 2, 1943, AB1/414, PRO.
40. Chadwick's views cited in Akers to Perrin, August 31, 1943, AB1/376, PRO; cf. JSM to WCO, October 23, 1943, AB1/376, PRO; W.A. Akers to Halban, January 1, 1943, AB1/414, PRO; Note by General Groves, July 22, 1943, *FRUS, Quebec Conference*, p. 635n.
41. Akers, "Negotiations with the Americans after the Signing of the Quebec Agreement," September 13, 1942, AB1/376, PRO; The "uneasy partnership" is well described in Richard G. Hewlett and Oscar E. Anderson, Jr., *The New World, 1939/1946*, Volume I of a *History of the United States Atomic Energy Commission* (University Park, Pennsylvania: The Pennsylvania State University Press, 1962), pp. 255–88.
42. V. Bush, "Memorandum for the President: Tube Alloys Interchange with the British," August 23, 1943, Atomic Bomb Folder, FDR Library; V. Bush, "Memorandum of Conference with Harry Hopkins and Lord Cherwell at the White House, May 25, 1943," Atomic Bomb Folder, FDR Library; Akers to Perrin, July 22, 1943, AB1/458, PRO; Richard G. Hewlett and Oscar E. Anderson, Jr., *A History of the United States Atomic Energy Commission*, Vol. I: *The New World, 1939–1946*, p. 271.
43. Andrew J. Pierre, *Nuclear Politics: The British Experience with an Independent Strategic Force, 1939–1970* (Oxford University Press, 1972), p. 5; cf. Marcus Oliphant, "The American Story," in *Atomic Challenge*, (London: Winchester Publications, 1947), p. 23.
44. Hewlett and Anderson, *The New World*, p. 179.
45. J.W. Pickersgill, *The Mackenzie King Record, 1939–1944*, Vol. I (Toronto: University of Toronto Press and University of Chicago Press, 1960), p. 532.
46. Bernstein, "The Uneasy Alliance," p. 214.
47. Gowing, *Independence and Deterrence*, Vol. I, p. 168.
48. Vannevar Bush, "Churchill and the Scientists," *Atlantic Monthly* Vol. CCXV (March 1965), p. 96.
49. Michael Armine, *The Great Decision: The Secret History of the Atomic Bomb* (New York: G.P. Putnam's 1959), pp. 125–6.
50. Hopkins Papers, cited in *Foreign Relations of the United States: Conference at Washington, 1941–1942, and Casablanca, 1943* (Washington: USGPO, 1968), 432.
51. Ibid., p. 803.
52. Churchill to Hopkins, February 27, 1943, *FRUS, Washington and Quebec*, p. 4.
53. *FRUS, The Conference in Washington and Quebec, 1943*, p. 633.
54. Illegible to M.W. Perrin, January 19, 1943, AB1/376, PRO.
55. Akers to Perrin, August 31, 1943, AB1/376, PRO.
56. B.L. Villa, "The Atomic Bomb and the Normandy Invasion," *Perspectives in American History*, Vol. 11 (1977/78), pp. 463–502.
57. *FRUS: The Conference at Washington and Quebec, 1943*, pp. 638, 1117.
58. Pierre, *Nuclear Politics*, pp. 20–1; A.J.R. Groom, *British Thinking About Nuclear Weapons* (London: Francis Pinter, 1974) pp. 6–7; George C. Laurence, "Canada's Participation in Atomic Energy Development," *Bulletin of the Atomic Scientists*, Vol. 3 (November 1947), pp. 325–9. The best overall study

is Martin J. Sherwin, *A World Destroyed: The Atomic Bomb and the Grand Alliance* (New York: Vintage, 1977); Winston S. Churchill, *The Hinge of Fate* (Boston: Houghton Mifflin, 1950), p. 809.

59. Bush, "Memorandum for the President," August 23, 1943; *FRUS: The Conference at Washington and Quebec,* p. 1097.

60. Pierre, *Nuclear Politics,* pp. 50–1. Cf. W.A. Akers to E.O. Lawrence, December 22, 1945, AB1/44, PRO.

61. Leslie R. Groves, *Now It Can Be Told: The Story of the Manhattan Project* (New York: Harper, 1962; 1983), p. 136.

62. Bundy, *Danger and Survival,* p. 108; Churchill, *The Hinge of Fate,* pp. 380–1.

63. Sir John Cockcroft, Review of *Britain and Atomic Energy,* by Margaret Gowing, *Disarmament and Arms Control,* Vol. III (Spring, 1965), p. 89.

64. Stephane Groueff, *Manhattan Project: The Untold Story of the Making of the Atomic Bomb* (New York: Bantam Books, 1968), pp. 304–8

65. James Kunetka, *City of Fire* (Albuquerque: University of New Mexico Press, 1982); Ferenc M. Szasz, *The Day the Sun Rose Twice: The Story of the Trinity Site Nuclear Explosion, July 16, 1945* (Albuquerque: University of New Mexico Press, 1984); John H. Manley, "Assembling the Wartime Labs," *Bulletin of the Atomic Scientists,* Vol. 30 (May 1974), pp. 42–7; Statement by Leo Szilard, US Congress, Senate Committee Hearings, 79th Congress, Vol. 809, 1946, Special, pp. 290–8.

66. Oppenheimer, cited in Nichols, *Road to Trinity,* p. 74; Szasz, *The Day the Sun Rose Twice,* pp. 16–17.

67. Phyllis K. Fisher, *Los Alamos Experience* (Tokyo and New York: Japan Publications, 1985), p. 39.

68. Ruth Marshak, "Secret City," in "The Atom and Eve," unpublished typescript, LANL Archives, p. 10; Goldschmidt, *The Atomic Adventure,* p. 26; Ray Powell, as cited in Furman, *A History of Sandia Laboratories,* p. 48.

CHAPTER 2   THE BRITISH MISSION AT LOS ALAMOS: THE SCIENTIFIC DIMENSION

1. The only studies of the British Mission at Los Alamos are two brief articles: Dennis C. Eakley, "The British Mission," *Los Alamos Science* (Winter/Spring 1983), pp. 186–9; and Pat Metropolis "The British Connection," *Los Alamos Newsbulletin* (February 11, 1983), pp. 8–9, David Hawkins has a good section on their technical work in *Toward Trinity,* Part 1 of *Project Y: The Los Alamos Story* (Los Angeles: Tomash, 1983).

2. Angus Calder, *The People's War: Britain, 1939–45* (London: Panther Books, 1971), pp. 318–20, 776–77, 439–42.

3. Otto Frisch, *What Little I Remember* (Cambridge: Cambridge University Press, 1979); French interview; M.J. Poole interview; D.J. Littler interview. Copies, LANL.

4. Rudolf Peierls, *Bird of Passage: Recollections of a Physicist* (Princeton: Princeton University Press, 1985), pp. 194–5.

5. French interview. French to author, August 18, 1989.

6. Interview with Carson Mark, September 1, 1987; from Air Ministry to TSM; for Webster from Perrin, February 1, 1944, AB1/481, PRO.
7. The Mark quotation is from review of Jordan Carson Mark. DF 65–58805, Vol. 31, Part II, Serials, pp. 1147–1186, Klaus Fuchs Papers; Hawkins, *Project Y*, p. 28.
8. Interview with John Manley, December 10, 1987; Edith C. Truslow and Ralph Carlisle Smith, *Beyond Trinity*, Part II of *Project Y: The Los Alamos Story*, Los Angeles: Tomash, 1983, pp. 26–9; W.L. Webster to Groves, March 16, 1944, AB1/481, PRO.
9. James W. Kunetka, *Oppenheimer: The Years of Risk* (Englewood Cliffs, NJ: Prentice Hall, 1982), p. 49.
10. R.C. Smith, "Report of Foreign Personnel at Project Y," June 3, 1944, LANL.
11. Lord Zuckerman, "Nuclear Wizards," *New York Review of Books*, March 31, 1988, p. 28; Richard Rhodes, *The Making of the Atomic Bomb* (New York: Simon and Schuster, 1986), pp. 543–6.
12. Oppenheimer to Groves, February 14, 1944. In Robert C. Williams and Philip L. Cantelon (eds), *The American Atom: A Documentary History of Nuclear Policies from the Discovery of Fission to the Present, 1939–1984* (Philadelphia: University of Pennsylvania Press, 1984), pp. 36–7.
13. Lewis Alvarez, *In the Matter of J. Robert Oppenheimer* (Cambridge, Mass.: MIT Press, 1971), p. 949.
14. Chuck Hansen, *U.S. Nuclear Weapons: The Secret History* (New York: Crown, 1987), p. 44.
15. Peierls, *Bird of Passage*, p. 200.
16. Cited in Frisch, *What Little I Remember*, p. 153.
17. Interviews with Mrs Hanni Bretscher, French and Poole.
18. French interview.
19. French to author, August 18, 1989.
20. *The Times*, June 30, 1975, 14e; see also October 15, 1962, 8g.
21. Smith to Bradbury, July 18, 1949, LANL Archives.
22. Cited in Metropolis, "The British Connection," *Los Alamos Newsbulletin*, pp. 8–9.
23. Margaret Gowing, *Britain and Atomic Energy, 1939–1945* (London: Macmillan, 1964), p. 265.
24. Barton C. Hacker, *The Dragon's Tail: Radiation Safety in the Manhattan Project, 1942–1946* (Berkeley: University of California Press, 1987), pp. 72–3.
25. Frisch, *What Little I Remember*, 160; Otto Frisch, "The Los Alamos Experience," *New Scientist*, Vol. 83 (July 19, 1979), p. 187; Kunetka, *Oppenheimer: The Years of Risk*, p. 65.
26. R.V. Jones, *The Wizard War: British Scientific Intelligence, 1939–1945* (New York: Coward, McCana & Geoghegan, 1978), p. 29; James L. Tuck, "Lord Cherwell and His Part in World War II," typescript, Los Alamos.
27. J.L. Tuck, "Autobiographical Notes," typescript, LANL Archives. cf. R.V. Jones, "Oxford Physics in Transition: 1929–1939," in Rajkumar Williamson (ed.), *The Making of Physicists* (Bristol: Adam Hilger, 1987).
28. Westfall interview with Peierls, September 19, 1986; Tuck, "Autobiographical Notes." Both in LANL, Archives.

29. Ibid.
30. Cf. Edward Teller, *Better a Shield Than a Sword: Perspectives on Defense and Technology* (New York: Free Press, 1987), p. 52.
31. Jones, *The Wizard War*, p. 29. Interview with P.B. Moon, September 23, 1987.
32. "Report on the Work by H. Sheard and D.J. Littler in Connection with the Atomic Bomb Test in New Mexico, 16th July 1945," AB1/414, PRO.
33. Marley to Sir James Chadwick, 28 September 1945, AB1/350, PRO.
34. Chadwick, cited in Margaret Gowing, "Reflections on Atomic Energy History," *The Bulletin of the Atomic Scientists*, Vol. 51 (March 1979), p. 52; interview with Sarah Tuck, August 30, 1989.
35. Morrison in Lansing Lamont Papers, HST Library.
36. R.C. Smith to J.L. Tuck, 29 January 1956, LANL Archives; Ralph Carlisle Smith to Norris Bradbury, July 18, 1949, LANL Archives.
37. Robert A. Lavender to Patent Group Heads, September 26, 1944, LANL Archives; R.C. Smith to Whitney Ashbridge, April 21, 1944, LANL Archives; cf. "Atomic Energy and U.S. Patent Policy," *Bulletin of the Atomic Scientists* (November 1, 1946), p. 28; Hewlett and Anderson, *The New World*, pp. 284–5.
38. Interview with Ralph Carlisle Smith by Robert Krohn, December 12, 1981. (T–81–0009), Copy LANL Archives, p. 14; Smith Orbituary, *Albuquerque Journal*, March 2, 1989.
39. Truslow and Smith, *Beyond Trinity*, Part II of *Project Y: The Los Alamos Story*, p. 62.
40. Robert A. Lavender to Ralph Carlisle Smith, March 17, 1945, Folder A–85–001, 1–1, LANL Archives; Smith to Lavender, March 7, 1945, LANL Archives; interview with Ralph Carlisle Smith, November 13, 1986, Copy, LANL Archives.
41. Interview with D.G. Marshall.
42. Winston S. Churchill, *Triumph and Tragedy* (Boston: Houghton Mifflin, 1950), p. 639.
43. Norman Ramsey to Groves, October 14, 1943; Arnold, as cited in Necah Stewart Furman, *A History of Sandia Laboratories – The Post-War Decade* (Albuquerque: University of Mexico Press, 1990), pp. 101, 189.
44. Martin Gilbert, *"Never Despair," Winston S. Churchill, 1945–1965* (London: Heinemann, 1958), p. 792.
45. Gilbert, *"Never Despair,"* pp. 59, 100–1.
46. Calder, *The People's War*, p. 676.
47. Gilbert, *"Never Despair,"* p. 249.
48. Truman, recollection for TV series, "The Conflicts of Harry S Truman," in Lansing Lamont Papers, Box 1, Research materials, HST Library; Address to Congress, HST Library.
49. Cf. Michael S. Sherry, *The Rise of American Air Power: The Creation of Armageddon.* (New Haven, Conn.: Yale University Press, 1987), pp. 117ff; Lee Kensett, *A History of Strategic Bombing* (New York: Scribner's, 1982).
50. Spencer Weart, *Nuclear Fear: A History of Images* (Cambridge, Mass.: Harvard University Press, 1988), p. 95.
51. Peierls' review of Rhodes, *The Making of the Atomic Bomb* in *New York Review of Books*, November 5, 1987, p. 48.

52. Lawrence in Lansing Lamont Papers, HST Library.
53. From: Oppenheimer to: Project Employees, May 4, 1945; *Bulletin*, February 7, 1944.
54. Arthur Compton, cited in "Tentative Chronology of Part Played by Scientists in Decision to Use the Bomb Against Japan," May 24, 1957, Papers of Harry S Truman, Post-Presidential files, HST Library; Churchill, *Triumph and Tragedy*, p. 639.
55. Robert J. Donovan, *Conflict and Crisis: The Presidency of Harry S Truman, 1945–1948* (New York: W.W. Norton, 1977), p. 66.
56. Littler interview; French interview; Marshall interview.
57. Clipping from the *Birmingham Evening Mail*, May 1986. Copy sent by Prof. W.N. Everitt.
58. Alice Kimball Smith, *A Peril and a Hope: The Scientists' Movement in America, 1945–47* (Cambridge, Mass.: MIT Press, 1965), p. 5.
59. *New York Times*, July 15, 1974 (Chadwick obituary).
60. Tizard, quoted in Ronald W. Clark, *Birth of the Bomb: The Untold Story of Britain's Part in the Weapon That Changed the World* (London: Phoenix House, 1961), p. 194.
61. See Paul Fussell, *Thank God for the Atomic Bomb and Other Issues* (New York: Summit, 1988). Interviews with William T. Hagan and Clarence Critzman.
62. Robert Porton interview.
63. Peierls, *Bird of Passage*, passim.
64. Interview with P.B. Moon.
65. Peggy Titterton interview.
66. J.G. Rushbrooke review of *What Little I Remember*, in *The Times Higher Education Supplement*, May 5, 1980.

CHAPTER 3   THE BRITISH MISSION AT LOS ALAMOS: THE SOCIAL DIMENSION

1. Interview with Sir Ernest Titterton, March 1989, Copy, LANL Archives; Rudolph Peierls, *Bird of Passage: Recollection of a Physicist* (Princeton: Princeton University Press, 1985), p.189; interview with D.G. Marshall, Copy, LANL Archives.
2. Interview with Sir Rudolf Peierls, January 13, 1988, Copy LANL Archives.
3. Laura Fermi, *Atoms in the Family: My Life with Enrico Fermi* (Chicago: University of Chicago Press), pp. 207–11; Peierls, *Bird of Passage*, p. 189; Ulam, as found in Lansing Lamont papers, Box 1, Research Material, Interviews, HST Library.
4. Interview (telephone) with Alice and Cyril Smith, Fall 1987, Copy LANL Archives.
5. Jean Bacher, "Fresh Air and Alcohol," in *The Atom and Eve*, unpublished MSS, LANL Archives.
6. Interview with M.J. Poole.
7. "Memorandum for Guidance of Technical Section of the British Supply Council Personnel," January 13, 1944, AB1/481, PRO; Paul Filipkowski,

"Postal Censorship at Los Alamos 1943–1945", *American Philatelist* (April 1987), p. 350.

8. Dorothy McKibben, "109 East Palace," *The Atom and Eve.*
9. Peggy Titterton interview.
10. Interview with Hanni Bretscher.
11. Otto Frisch, *What Little I Remember* (Cambridge: Cambridge University Press, 1929).
12. Interviews with John Manley and Alice Cyril Smith.
13. Peggy Titterton interview.
14. Interview with Joseph Rotblat, January 13, 1988.
15. Peierls interview.
16. Frisch, *What Little I Remember*, p. 158; cf. J.H. Manley, "Assembling the Wartime Labs," *Bulletin of the Atomic Scientists*, Vol. 30 (1974).
17. Interview with John Manley.
18. Felix Diaz Almaraz, Jr., "The Little Theatre in the Atomic Age: Amateur Dramatics in Los Alamos, New Mexico, 1943–1946," *Journal of the West*, Vol. 17 (1978).
19. Data from Los Alamos Daily News Bulletins, lent by Robert Porton.
20. Interview with J. Carson Mark.
21. Interview with Alice and Cyril Smith.
22. Ruth Marshak, "Secret City," in *The Atom and Eve.*
23. Bernice Brode, "Tales of Los Alamos," *LASL Community News*, August 11, 1960, p. 7.
24. Interview with Sarah Tuck.
25. Peggy Titterton interview.
26. Interview (telephone) with Sir Rudolf Peierls, January 13, 1988. Copy LANL Archives.
27. Anecdote from Kate Blewett, who heard it from Mrs Norris Bradbury, Spring 1989.
28. Memorandum: Re: Rudolf Ernst Peierls, January 24, 1951, Fuchs file 65–58805, Vol 41. Serials 1457–1500; Fermi, *Atoms in the Family*, p. 208.
29. Manley interview.
30. Porton interview; Tuck "Autobiographical Notes." Typescript, LANL Archives.
31. J. Carson Mark interview; Hugh Jennings, "Lab Experiences," MSS, LANL Archives; Manley interview.
32. Americans requested that Frisch be naturalized before he went to Los Alamos. Richard Tolman to Akers, November 10, 1943, AB1/376, PRO.
33. Brode, "Tales of Los Alamos"; interview with Mrs Hanni Bretscher.
34. Source prefers to remain anonymous.
35. Bretscher interview.
36. Margaret Gowing, "Britain, America and the Bomb," in David Dilks (ed.), *Retreat from Power: Studies in Britain's Foreign Policy of the Twentieth Century*, Volume II: *After 1939* (London: Macmillan, 1981), p. 128.
37. Mark Oliphant, "James Chadwick," *Physics Today*, Vol. 276 (October 1974), pp. 87–8.
38. Interview with Ernest Titterton.
39. Chadwick to Peierls, April 6, 1945, AB1/485.
40. Peierls, *Bird of Passage*, p. 201.

41. Bretscher interview.
42. Cited in Margaret Gowing, *Britain and Atomic Energy, 1939–45* (London: Macmillan, 1964), p. 262.
43. Titterton interview.
44. Tuck, *Autobiographical Notes*. Typescript, LANL Archives.
45. Otto Frisch, *What Little I Remember*, p. 176.
46. Ralph Carlisle Smith interview.
47. Smith interview; Frisch, *What Little I Remember*, pp. 175–6.
48. Interview with Mrs Hanni Bretscher.
49. Smith interview. Only bachelor Otto Frisch recalled having adequate funds, *What Little I Remember*, p. 156; cf. P.B. Moon to R.C. Smith, November 4, 1944, LANL Archives.
50. Kathleen Mark, "A Roof Over Our Heads," in *The Atom and Eve*.
51. Skyrme clipping, sent by Prof. W.N. Everitt.
52. Tuck, *Autobiographical Notes*, Typescript, LANL Archives, p. 6; Mark, "A Roof Over Our Heads."
53. Bretscher interview.
54. Peggy Titterton interview.
55. Interview with M.J. Poole.
56. Peierls, *Bird of Passage*, p. 206; details from Bretscher interview; Pat Metropolis, "The British Connection, *News Bulletin*, February 11, 1983, pp. 8–9; and Mrs Bernice Brode, "Life at Los Alamos, 1943-45," *Atomic Scientists Journal*, Vol. 3 (November 1953), pp. 87–91; Titterton interview. See also Jane S. Wilson and Charlotte Serber (eds), *Standing By and Making Do: Women of Wartime Los Alamos* (Los Alamos Historical Society, 1988), p. 114.
57. Brode, "Tales of Los Alamos," pp. 7–8; Dennis C. Fakley, "The British Mission," *Los Alamos Science*, Vol. IV (Winter/Spring 1983), pp. 188–9; Smith interview; Titterton interview.
58. Titterton interview.
59. Ibid.
60. Smith, "Report of Foreign Personnel at Project Y," February 5, 1946, LANL Archives.
61. J. Chadwick to P. Moon, September 10, 1945, AB1/485, PRO; E.W. Titterton to Sir James Chadwick, 27 September 1945, AB1/445, PRO.
62. Sir John Cockcroft, Review of Margaret Gowing, "*Britain and Atomic Energy*" in *Disarmament and Arms Control*, Vol. III (Spring, 1963) p. 89.
63. Quoted in Peter Malone, *The British Nuclear Deterrent* (New York: St. Martin's Press, 1984), p. 3.
64. Norris Bradbury in *Reminiscences of Los Alamos, 1943–1945* (Dordrecht, Netherlands: Reidel, 1980), p. 169–70.
65. The phrase is Christopher Thorne's, as found in Alex Danchev, *Very Special Relationship* (London: Brassey's Defence Publishers, 1986), p. 133.

CHAPTER 4   THE AFTERMATH

1. This period is expertly detailed in Alice Kimball Smith, *A Peril and a Hope:*

The Scientists' Movement in America: 1945–47 (Cambridge, Mass.: MIT Press, 1965; 1970). See also Byron S. Miller, "A Law Is Passed – The Atomic Energy Act of 1946," *The University of Chicago Law Review*, Vol. 15 (Summer 1948), pp. 799–821; Gregg Herken, *Counsels of War* (New York: Knopf, 1985) tells of the advisors who influenced American nuclear policy in the postwar period.

2. Forrestal, as cited in Robert J. Donovan, *Conflict and Crisis: The Presidency of Harry S Truman, 1945–1948* (New York: W. W. Norton, 1977), p. 143; Raymond Gram Swing to Ross, September 29, 1945, Papers of Harry Truman, Official File, HST Library; Necah Stewart Furman, *The History of Sandia Laboratories – The Postwar Decade* (Albuquerque: University of New Mexico Press, 1990), p. 214.

3. Interview with A.P. French.

4. G.A. McMillan to E.W. Titterton, June 19, 1946, AB1/445, PRO; Peggy Pond Church, *House at Otowi Bridge* (Albuquerque: University of New Mexico Press), p. 19, tells the story of Edith Warner and the Los Alamos scientists.

5. "Atomic Energy Agreed Declaration," President's Secretaries' Files, HST Library.

6. Lloyd J. Graybar, "The 1946 Atomic Bomb Tests: Atomic Diplomacy or Bureaucratic Infighting?" *Journal of American History*, Vol. 72 (March 1986), p. 901.

7. Tuck, *Autobiographical Notes,* Typescript, LANL Archives; Titterton interview.

8. W.A. Shurcliff, *Bombs at Bikini: The Official Report of Operation Crossroads* (New York: Wm. H. Wise, 1947).

9. See Richard G. Hewlett and Oscar E. Anderson, *A History of the United States Atomic Energy Commission,* Vol. I: *The New World, 1939/1946* (University Park, Penn.: Pennsylvania State University Press, 1962), for the best discussion of this. Corbin Allardice to R.C. Smith, November 7, 1947; R.C. Smith, "Review of British Crossroads Report," November 26, 1947, LANL Archives.

10. November 11, 1945 Statement, in President's Secretary's Files, HST Library.

11. Smith interview.

12. Carson Mark to N.E. Bradbury, November 7, 1946, LANL Archives.

13. Groves to Bradbury, August 14, 1946; Mark to Bradbury, November 7, 1946; Mark interview.

14. R.C. Smith, "Report of Foreign Personnel at Project Y," 1 April 1947, LANL Archives; H.C. Gee to Roger S. Warner, "Report of Foreign Personnel at Project Y," April 7, 1947, LANL Archives; H.C. Gee to Carroll L. Wilson, 2 May 1947, LANL Archives.

15. James L. Gormly, "The Washington Declaration and the 'Poor Relation': Anglo-American Atomic Diplomacy, 1945–46," *Diplomatic History*, Vol. 81 (Spring 1984), pp. 125–43. Hewlett and Anderson, *The New World, 1939/1946*, pp. 478–81.

16. Eduard Mark, " 'Today Has Been a Historical One': Harry S Truman's Diary of the Potsdam Conference," *Diplomatic History*, Vol. 4 (1980), p. 320; Robert H. Pilpel, *Churchill in America, 1895–1961: An Affectionate Portrait* (New York: Harcourt, Brace Jovanovich, 1976), p. 250.

17. "Minutes of Meeting of the Secretaries of State, War, and the Army,"

October 16, 1945, *FRUS: Diplomatic Papers*, Vol. II, General, p. 59; Anderson to Truman, September 25, 1945. President's Secretary's Files, HST Library. McKellar to Truman, September 27, 1945. Ibid.

18. Anderson to Truman, September 25, 1945, HST Library.

19. Truman as quoted in *Life* (August 20, 1945), p. 32; and in A.J.R. Groom, *British Thinking About Atomic Weapons* (London: Francis Pinter, 1974), p. 25.

20. Francis Duncan, "Atomic Energy and Anglo-American Relations, 1946–1954," *Orbis*, Vol. 12 (Fall 1968), pp. 1190–1.

21. David E. Lilienthal, *The Journals of David E. Lilienthal*, Vol. II: *The Atomic Energy Years, 1945–1950* (New York: Harper & Row, 1964), pp. 175–6.

22. Quoted in Groom, *British Thinking About Atomic Weapons*, p. 29.

23. Cf. the estimates by Frederick Seitz and Hans Bethe, and those by Irving Langmuir, both in *One World or None* (New York: McGraw Hill, 1946), pp. 42–6; 49; *New York Times*, August 13, 1945.

24. Personal Diary of Charles G. Ross, 1946; Papers of Charles G. Ross, Box 21, HST Library, Cf. Lilienthal, *Atomic Energy Years*, p. 636.

25. Gregg Herken, "'A Most Deadly Illusion': The Atomic Secret and American Nuclear Weapons Policy, 1945–1950," *Pacific Historical Review*, Vol. XLIX (February 1980), pp. 51–77; David Alan Rosenberg, "A Smoking Radiating Ruin at the End of Two Hours: Documents on American Plans for Nuclear War with the Soviet Union, 1954–1955," *International Security*, Vol. 6 (Winter 1981–82), pp. 1–37; David Alan Rosenberg, "U.S. Nuclear Stockpile, 1945–1950," *Bulletin of the Atomic Scientists*, Vol. 38 (May 1982), pp. 25–30.

26. Attlee to Truman, 7 June 1946, President's Secretary's Files, HST Library; Memorandum by the Secretary of War (Stimson) to President Truman, September 11, 1945, *FRUS:* Diplomatic Papers, Vol. II, General, p. 41.

27. Duncan, "Atomic Energy and Anglo-American Relations, 1946–1954," pp. 1197–8; Timothy J. Botti, *The Long Wait: The Forging of the Anglo-American Nuclear Alliance, 1945–1958* (New York: Greenwood, 1987), p. 2.

28. Richard G. Hewlett and Frances Duncan, *A History of the United States Atomic Energy Commission*, Vol. II: *Atomic Shield, 1947–1952* University Park, Penn. and London: Pennsylvania State University Press, 1969); Cf. Michael Howard's Review of Gowing in *The Times* December 8, 1974, p. 38c.

29. Luis Alvarez, *Alvarez: Adventures of a Physicist* (New York: Basic Books, 1987), pp. 127–8.

30. Acheson, cited in Duncan, "Atomic Energy and Anglo-American Relations, 1946–1954," p. 1202.

31. Alfred Goldberg, "The Atomic Origins of the British Nuclear Deterrent," *International Affairs*, Vol. 40 (July 1964), p. 418.

32. Randall Bennett Woods, *A Changing of the Guard: Anglo-American Relations, 1941–1946* (Chapel Hill: University of North Carolina Press, 1990), pp. 405–6.

33. Chadwick, quoted in Margaret Gowing, "Britain, America and the Bomb," in David Dilks (ed.), *Retreat from Power: Studies in Britain's Foreign Policy of the Twentieth Century*, Vol. II: *After 1939* (London: Macmillan, 1981), p. 132.

34. Westfall interview with Peierls, September 19, 1986, LANL Archives; Margaret Gowing, *Independence and Deterrence: Britain and Atomic Energy, 1945–1953*, Vol. I: *Policy Making* (London: Macmillan, 1974), p. 52.

35. Harold Macmillan, *Riding the Storm, 1956–1959* (New York: Harper & Row, 1971), p. 320; Richard G. Hewlett and Jack M. Hall, *Atoms for Peace and War, 1953–1961: Eisenhower and the Atomic Energy Commission* (Berkeley: University of California Press, 1989), p. 467.

36. Macmillan, *Riding the Storm*, pp. 316–23.

37. Macmillan, *Riding the Storm*, p. 319.

38. Macmillan, *Riding the Storm*, p. 323.

39. Lorna Arnold, *A Very Special Relationship: British Atomic Weapons Trials in Australia* (London: HMSO, 1987), p. 86; John Simpson, *The Independent Nuclear State: The United States, Britain and the Military Atom* (New York: St. Martin's Press, 1933), p. 143.

40. Alistair Horne, *Harold Macmillan*, Vol. II, *1957–1986* (New York: Viking, 1989), pp. 53–4.

41. Simpson, *Independent Nuclear State*, p. xxix.

42. Woods, *A Changing of the Guard*, p. 1.

43. David Reynolds, "Roosevelt, Churchill, and the Wartime Anglo-American Alliance, 1939–1945: Towards a New Synthesis," in William Roger Louis and Hedley Bull (eds), *The 'Special Relationship' Anglo-American Relations Since 1945* (New York: Oxford, 1986), pp. 18–19; Henry Butterfield Ryan, *The Vision of Anglo-America: The US–UK Alliance and the Emerging Cold War, 1943–1946* (Cambridge: Cambridge University Press, 1987).

CHAPTER 5   VARIETIES OF THE BRITISH MISSION EXPERIENCE

1. Joseph Rotblat, "Leaving the Bomb Project," *Bulletin of the Atomic Scientists*, Vol. 41 (August 1985), pp. 16–19.

2. Rotblat in *The Times*, November 6, 1981, pp. 15f.

3. J. Jerome Maxwell report 1/23/50; Fuchs File No. 65–58805; Vol. No. 2. Serials, pp. 27, 82.

4. Interview with Joseph Rotblat, January 13, 1988. Copy LANL Archives.

5. "Why I Stopped Work on the Bomb," *The Times*, July 17, 1985, p. 10a.

6. Rotblat interview.

7. It is not clear how much knowledge Los Alamos had of the success or failure of the German atomic project in late 1944. See Samuel A. Goudsmit, *Alsos* (New York: Henry Schuman, 1947), Boris T. Pash, *The Alsos Mission* (New York: Award House, 1965), and Mark Walker, *German National Socialism and the Quest for Nuclear Power* (New York: Cambridge University Press, 1989).

8. *New York Times*, August 10, 1945; *Time* (August 20, 1945); Robert Jungk, *Brighter than a Thousand Suns* (New York: Harcourt, Brace, 1958), p. 109; Robert Jungk, "Los Alamos – Life in the Shadow of the Atomic Bomb," typescript, Ralph Carlisle Smith Collections, Coronado Room, Zimmerman Library, University of New Mexico.

9. Bloch, quoted in John Rigden, *Rabi: Scientist and Citizen* (New York: Basic

Books, 1987), p. 153.

10. Conversation with Norris Bradbury, spring 1985. Bloch seldom talked about his Los Alamos experiences afterwards.

11. Report by J. Jerome Maxwell, 5/3/50, KF File 65–58805, Vol. 29, Serials 1106 to 1145. Another version is in J. Jerome Maxwell 1/23/50 File 65–58805, Vol. 2, Serials, 27–82.

12. J. Jerome Maxwell, "Emil Julius Klaus Fuchs," May 3, 1950, KF 65–58805, Vol. 29, Serials 1106–1145.

13. To Director from: SAC, San Francisco Subject: Foocase. Espionage – R, November 7, 1950, File No. 65–58805, Vol. 41, Serials, 1457 through 1500.

14. Bretscher interview. Egon Bretscher and Rotblat shared offices at Los Alamos.

15. Joseph Rotblat, "Leaving the Bomb Project," *Bulletin of the Atomic Scientists*, Vol. 41 (August 1985), pp. 16–19.

16. *The Times*, September 19, 1983, p. 6a.

17. J. Rotblat, "The Hydrogen-Uranium Bomb," *Bulletin of the Atomic Scientists*, Vol. 11 (May, 1955), pp. 171–2, 177; Robert A. Divine, *Blowing on the Wind: The Nuclear Test Ban Debate, 1954–1960* (New York: Oxford, 1978), pp. 48–9.

18. *The Times*, October 28, 1955, p. 9e; see also *The Sunday Times*, January 16, 1977, 2f; ibid., October 9, 1976, p. 12g; ibid., April 13, 1977, p. 13d.

19. *The Times*, October 4, 1983, p. 2d.

20. *The Times Higher Education Supplement*, June 14, 1977, p. 18g.

21. *The Times*, June 6, 1972, p. 15a.

22. Joseph Rotblat, "British Fret about 'Vulnerability,'" *Bulletin of the Atomic Scientists*, Vol. 44 (March 1984), p. 22.

23. Rotblat, in *The Times Higher Education Supplement*, December 20, 1981, p. 18.

24. L.G. Brookes, in *The Times Higher Education Supplement,* December 18, 1981, p. 23c.

25. Rotblat interview; "Why I Stopped Work on the Bomb," *The Times*, July 17, 1985, p. 10a.

26. July 1988 Statements by Lord William George Penney to written questions posed by Ferenc M. Szasz. Copy, LANL Archives. The only biographical sketch of Penney is Margaret Gowing's excellent summation in "The Men," Chapter 13 of *Independence and Deterrence: Britain and Atomic Energy, 1945–1953*, Vol. II: *Policy Execution* (London: Macmillan, 1974), pp. 2–23. Lorna Arnold also has a good bit on his role in *A Very Special Relationship: British Atomic Weapons Trials in Australia* (London: HMSO, 1987), *passim*.

27. Peter Wyden, *Day One: Before Hiroshima and After* (New York: Simon & Schuster, 1984), p. 194n.

28. Richard G. Hewlett and Oscar E. Anderson, Jr., *A History of the United States Atomic Energy Commission*, Vol. I: *The New World, 1939/1946* (University Park: Pennsylvania State University Press, 1962), p. 377.

29. Penney in Lansing Lamont Papers, HST Library.

30. July 1988, Statements by Lord William George Penney; Penney statement in Lansing Lamont Papers, HST Library.

31. Quoted in Richard Rhodes, *The Making of the Atomic Bomb* (New York: Simon & Schuster, 1986), p. 678.

32. Wyden, *Day One*, p. 194.

33. Vincent C. Jones, *Manhattan: The Army and the Atomic Bomb* (Washington: USGPO, 1984), p. 528.
34. Wyden, *Day One*, p. 194.
35. July 1988, Statements by Lord William George Penney. N.E. Bradbury Certificate, 7 May 1946, LANL Archives.
36. Leslie R. Groves, *Now It Can Be Told: The Story of the Manhattan Project* (New York: Da Capo Press, 1962; 1983), p. 282n.
37. Luis W. Alvarez, *Alvarez: Adventures of a Physicist* (New York: Basic Books, 1987), p. 144. See also, W. Peter Trower (ed.), *Discovering Alvarez: Selected Works of Luis W. Alvarez, with Commentary by His Students and Colleagues* (Chicago: University of Chicago Press, 1987), pp. 55–71.
38. Brian Gardner, *The Year That Changed the World, 1945* (New York: Coward-McCann, 1962), pp. 261–2.
39. Gowing, *Independence and Deterrence*, Vol. II, p. 7.
40. ibid., p. 6.
41. Arnold, *A Very Special Relationship*, p. 14; Gowing, *Independence and Deterrence*, Vol. II, p. 11; Cherwell quoted in Arnold, p. 14.
42. *The Times*, December 8, 1974, p. 38c.
43. Margaret Gowing, "Nuclear Weapons and the 'Special Relationship'," in William Roger Louis and Hedley Bull (eds), *The Special Relationship: Anglo-American Relations since 1945* (New York: Oxford, 1986), p. 120.
44. Penney, quoted in Michael Howard, "The Explosive Secret," *The Times*, December 8, 1974, p. 38c.
45. Margaret Gowing, *Independence and Deterrence*, Vol. II: *Policy Making*, p. 213.
46. Peter Malone, *The British Nuclear Deterrent* (New York: St. Martin's Press, 1984), p. 6.
47. Gowing, *Independence and Deterrence*, Vol. I, p. 52.
48. Arnold, *A Very Special Relationship*, p. 228 and *passim*. The quotation is from Peter Hennessy, *Whitehall* (London: Secker & Warburg, 1989), p. 713.
49. *The Times*, May 15, 1952, p. 6e; October 15, 1953, p. 8e.
50. *The Times*, October 3, 1952, p. 6a; J.L. Symonds, *A History of British Atomic Tests in Australia* (Canberra: Australia Government Publishing Service, 1985), p. 67–8.
51. John Simpson, *The Independent Nuclear State: The United States, Britain, and the Military Atom* (New York: St. Martin's Press, 1983), pp. 73–4.
52. Simpson, *The Independent Nuclear State*, p. xviii.
53. Gowing, *Independence and Deterrence*, Vol. II, p. 449; Vol. II, p. 18.
54. *The Times*, October 24, 1952, p. 8b.
55. *The Times*, October 3, 1952, p. 6b; October 4, 1952, p. 4d. One month later, the US detonated its first hydrogen bomb; the Russians followed in 1955.
56. *The Times*, October 24, 1952, p. 8b.
57. *The Times*, September 25, 1953, p. 8f.
58. Arnold, *A Very Special Relationship*, p. 47.
59. Arnold, *A Very Special Relationship*, pp. 73–5; Symonds, *A History of British Atomic Tests in Australia*, p. 193; Robert Milliken, *No Conceivable Injury: The Study of Britain and Australia's Atomic Cover-up* (Ringwood, Victoria, Australia: Penguin, 1986), pp. 121–30.
60. Arnold, *A Very Special Relationship*, p. 183.

61. Richard G. Hewlett and Jack M. Holl, *Atoms for Peace and War, 1953–1961: Eisenhower and the Atomic Energy Commission* (Berkeley: University of California Press, 1989), p. 548.
62. Timothy J. Botti, *The Long Wait: The Forging of the Anglo-American Nuclear Alliance, 1945–1958* (New York: Greenwood, 1987).
63. Hewlett and Holl, *Atoms for Peace and War*, pp. 538–9.
64. *The Times*, September 21, 1965, p. 8a.
65. *The Times*, October 18, 1966, p. 93; February 19, 1966, p. 9c.
66. *The Times*, September 24, 1965, p. 10f.
67. *The Times*, July 9, 1963, p. 9b; July 19, 1963, p. 10b; July 20, 1963, p. 8b.
68. *The Times*, October 25, 1968, p. 3c.
69. Arnold, *A Very Special Relationship*, p. 219.
70. All Titterton material comes from the several-hour reminiscence given by Titterton in March 1989. Copy, LANL Archives.
71. *The Times*, May 22, 1985, p. 7a.
72. *The Times*, May 15, 1985, p. 7a.
73. "A Man of the Century," *Time*, Vol. LXXX (November 30, 1962), pp. 56–7; *The Times*, November 26, 1962, p. 12b.
74. *The Times*, November 19, 1962, p. 12c; "A Man of the Century," pp. 56–7; Frisch, as cited in Norman Moss, *The Politics of Uranium* (New York: Universe Books, 1981), p. 25.
75. Cited in John Manley in a review of Niels Bladel, *Harmony and Unity: The Life of Niels Bohr* (Madison, Wis: Science Tech, 1988), in the Albuquerque *Journal*, November 6, 1988.
76. Cited in Victor F. Weisskopf, "Niels Bohr, The Quantum, and the World," in A.P. French and P.J. Kennedy (eds), *Niels Bohr: A Centenary Volume* (Cambridge Mass.: Harvard University Press, 1985), p. 22.
77. Franck, quoted in *The Times Higher Education Supplement*, October 14, 1977, p. 20d.
78. James Franck, "A Personal Memoir," in French and Kennedy (eds), *Niels Bohr*, quoted on p. 17.
79. Victor F. Weisskopf, *The Privilege of Being a Physicist* (New York: W.H. Freeman, 1989), pp. 202–3.
80. *New York Times*, August 7, 1945.
81. Ruth Moore, *Niels Bohr: The Man, His Science, and the World They Changed* (Cambridge: MIT Press, 1985), p. 316.
82. The phrase is Margaret Gowing's. Margaret Gowing, "Niels Bohr and Nuclear Weapons," in French and Kennedy (eds), *Niels Bohr*, p. 269.
83. R.I. Campbell to J. Chadwick, December 28, 1943, AB1/No. 40, PRO.
84. Frisch, *What Little I Remember*, pp. 68–9; Necah Stewart Furman, *A History of Sandia Laboratories – The Postwar Decade* (Albuquerque: University of New Mexico Press, 1990), p. 84.
85. Oppenheimer to Groves, January 17, 1944. In Robert C. Williams and Philip L. Cantelon (eds), *The American Atom: A Documentary History of Nuclear Policies from the Discovery of Fission to the Present, 1939–1984* (Philadelphia: University of Pennsylvania Press, 1984), pp. 35–6.
86. Hans A. Bethe, "Niels Bohr and His Institute," in French and Kennedy (eds), *Niels Bohr*, p. 233.
87. Statement by Lord William George Penney.

88. Joseph O. Hirschfelder, "The Scientific and Technological Miracle at Los Alamos," in Lawrence Badash *et al.*, *Reminiscences of Los Alamos, 1943–1945* (Dordrecht, Netherlands: D. Reidel, 1980), p. 80; Philip Morrison statement, Lansing Lamont Papers, HST Library.

89. Chadwick in Lansing Lamont Papers, Box 1, Folder III, HST Library.

90. [J.] Robert Oppenheimer, "Niels Bohr and Atomic Weapons," *New York Review of Books* (December 17, 1964), p. 7; cf. Robert R. Wilson, "Niels Bohr and the Young Scientists," *Bulletin of the Atomic Scientists*, Vol. 41 (August 1985), pp. 23–6; Smith to Author, October 18, 1987.

91. Lansing Lamont Papers, HST Library.

92. J. Rud Nielsen, "Niels Bohr," *Physics Today*, Vol. 16 (October 1963), quoted on p. 28.

93. Margaret Gowing, *Britain and Atomic Energy, 1939–1945*, p. 246–50, 347–66; Gowing, "Niels Bohr and Nuclear Weapons," in French and Kennedy (eds), *Niels Bohr*, p. 26; see also Richard Rhodes, *The Making of the Atomic Bomb* (New York: Simon & Schuster, 1986), pp. 527–30.

94. Mark Oliphant, "Bohr and Rutherford," in French and Kennedy (eds), *Niels Bohr*, p. 68. Rudolf Peierls, *Bird of Passage: Recollections of a Physicist* (Princeton: Princeton University Press, 1985), p. 56.

95. Oliphant, "Bohr and Rutherford," p. 68; Stewart Cockburn and David Ellyard, *Oliphant: The Life and Times of Sir Mark Oliphant* (Adelaide: Axiom Books, 1981), p. 36.

96. Niels Bohr, "Natural Philosophy and Human Cultures," in Bohr, *The Philosophical Writings of Niels Bohr*, Vol. II, *Essays 1932–1957 on Atomic Physics and Human Knowledge* (Woodbridge, Conn.: Ox Bow Press, 1987), p. 31.

97. Interview with Ralph Carlisle Smith.

98. Interview with John Manley. Manley review of Niels Blaedel, *Harmony and Unity: the Life of Niels Bohr* (Madison, Wis.: Science Tech, 1988), in the Albuquerque *Journal*, November 6, 1988; interview with Donald Marshall.

99. J. Robert Oppenheimer, "Niels Bohr and Atomic Weapons," HST Library.

100. Rudolf Peierls, "Some Recollections of Bohr," in French and Kennedy (eds), *Niels Bohr*, pp. 229–30.

101. London *Times*, November 26, 1962, p. 12b.

102. Aage Bohr, "The War Years and the Prospects Raised by Atomic Weapons," in S. Rozental (ed.), *Niels Bohr* (Amsterdam: North-Holland, 1967), p. 204.

103. Cited in Margaret Gowing, "Niels Bohr and Nuclear Weapons," in French and Kennedy (eds), *Niels Bohr*, p. 271.

104. Quoted in R.V. Jones, "Meetings in Wartime and After," in French and Kennedy (eds), *Niels Bohr*, p. 285.

105. In 1954, Churchill remarked that the tragedy was that the West did not tell the Soviet Union everything it knew about the atomic bomb when the United States still had the opportunity. Philip Knightley, *The Second Oldest Profession: Spies and Spying in the Twentieth Century* (New York: W. W. Norton), p. 266n.

106. "Tube Alloys Aide-Mémoire of conversation between the President and the Prime Minister at Hyde Park, September 18, 1944," Atomic Bomb Folder, FDR Library; *Science*, Vol. 194 (October 8, 1976), p. 175.

107. Niels Bohr, "The Rutherford Memorial Lecture, 1958," and "The Genesis of

Quantum Mechanics," in Bohr, *The Philosophical Writings of Niels Bohr*, Vol. III, *Essays 1958–1962 on Atomic Physics and Human Knowledge* (Woodbridge, Conn.: Ox Bow Press, 1987), pp. 54, 78.

108. *Science*, Vol. 174 (October 8, 1976), p. 175.

109. Philip Morrison, "A Glimpse of the Other Side," in French and Kennedy (eds), *Niels Bohr*, p. 345.

110. Gowing, "Niels Bohr and Atomic Weapons," pp. 274–5.

111. *One World or None* (New York: McGraw-Hill, 1946); N.H.D. Bohr, "To the Advisory Committee on Atomic Energy," 27 August 1945. AB1/40, PRO.

112. Niels Bohr, "For an Open World," *Bulletin of the Atomic Scientists*, Vol. I (July 1950), p. 215.

113. Bohr, "For an Open World," p. 213.

114. John Archibald Wheeler, "Niels Bohr and Nuclear Physics," *Physics Today* (October 1963), p.44.

115. *The Times*, June 13, 1950; ibid., June 21, 1950, p. 3f.

116. "Remarks of Aage Bohr at Niels Bohr Memorial Session," *Physics Today* (October 1963), p. 33; see also, Aage Bohr, "The War Years and the Prospects Raised by the Atomic Weapons," in S. Rozental (ed.), *Niels Bohr*.

CHAPTER 6   THE STRANGE TALE OF KLAUS FUCHS

1. The best studies of Fuchs are the two new biographies, Norman Moss, *Klaus Fuchs: A Biography* (New York: St. Martin's Press, 1987) and Robert Chadwell Williams, *Klaus Fuchs, Atom Spy* (Cambridge, Mass., and London: Harvard University Press, 1987). Other accounts that deal with the theme are Chapman Pincher, *Too Secret Too Long* (New York: St. Martin's Press, 1984); Oliver Pilat, *The Atom Spies* (New York: G.P. Putnam's, 1952); H. Montgomery Hyde, *The Atom Bomb Spies* (London: Hamish Hamilton, 1980); Rebecca West, *The New Meaning of Treason* (New York: Viking, 1964; 1985); Alan Moorehead, *The Traitors* (New York: Harper & Row, 1952; 1963). But see also Eric M. Braindel, "Do Spies Matter?" *Commentary*, Vol. 85 (March 1988), pp. 53–8. A good summary of the Fuchs case may also be found in Margaret Gowing, *Independence and Deterrence: Britain and Atomic Energy, 1945–1953*, Vol II: *Policy Execution* (London: Macmillan, 1974), pp. 144–53.

2. "The Housewife Who Spied for the Russians," *The Sunday Times*, January 27, 1980, p. 13g. A booklet on her, *Sonia's Rapport*, appeared in East Germany in 1977; Kim Philby, *My Silent War* (New York: Ballantine, 1968).

3. Robert J. Lamphere and Tom Shachtman, *The FBI–KGB War: A Special Agent's Story* (New York: Random House, 1986), pp. 85–6, 133.

4. Hoover to Brian McMahon, April 6, 1950, KF, 65–58805, Vol. 28, Serials 1039–1105.

5. *Daily Telegraph*, July 23, 1979, p. 10c; *Daily Telegraph*, August 1, 1979, p. 3e; Richard Deacon is the pen name of Donald McCormack.

6. D.M. Ladd to the Director, February 13, 1950, KF, 65–58805, Vol. 6, Serials 301–385.

7. Williams, *Klaus Fuchs, Atom Spy*; J.E. Hoover to Sidney W. Sowers, March

7, 1950, HST Library; Nigel West, *The Circus: MI5 Operations, 1945–1972* (New York: Stein & Day, 1982), pp. 45–9.

8. D.M. Ladd to the Director, February 2, 1950, KF, 65–58805, Vol. 3, Serials 83–171.

9. Memo, Foocase, May 8, 1950, KF, 65–58805, Vol. 29, Serials 1106–1145.

10. The "controlled schizophrenia" quotation is found in many places. See H. Montgomery Hyde, *The Atom Bomb Spies*, p. 97.

11. J. Edgar Hoover, "The Crime of the Century," *Reader's Digest*, Vol. 58 (January–June 1951), pp. 150–7. Hoover elaborated on the story of Soviet espionage in his *Masters of Deceit* (New York: Henry Holt, 1958).

12. Memo: May 3, 1950, KF, 65–58805, Vol. 29, Serials 1106–1145.

13. *Manchester Evening News*, March 1, 1950; *Manchester Guardian*, March 2, 1950.

14. Max Born, *My Life: Recollections of a Nobel Laureate* (London: Taylor & Francis, 1978).

15. Williams, *Klaus Fuchs, Atom Spy*. Moss, *Klaus Fuchs: A Biography*.

16. Norman Moss in the *New York Times*, January 29, 1988, p. 13.

17. *New York Times*, March 2, 1950, p. 2f.

18. Richard G. Hewlett and Francis Duncan, *Atomic Shield, 1947/1952: A History of the United States Atomic Energy Commission* (University Park, Penn.: Pennsylvania State University Press, 1969), pp. 312–14.

19. A.H. Belmont to Mr Ladd, February 28, 1951, KF, 65–58805, Vol. 41, Serials 1457–1500.

20. Moss, *Klaus Fuchs*, p. 186.

21. Richard L. Garwin to Harry S Truman, December 11, 1952, HST Library.

22. Hoover, "The Crime of the Century," pp. 150–7.

23. *Bulletin of the Atomic Scientists* (April 1949), cited in Re: Rudolf Ernst Peierls, KF.

24. Radio Script for Congressman J. Glenn Beall, KF, 65–58805, Vol. 8, Serials 466–75.

25. Richard Gid Powers, *Secrecy and Power: The Life of J. Edgar Hoover* (New York: Free Press, 1987), pp. 300–5. See Also Ronald Radosh and Joyce Milton, *The Rosenberg File: A Search for the Truth* (New York: Vintage, 1984). They conclude that Julius was guilty but that Ethel was only an accessory.

26. Roger M. Anders (ed.), *Forging the Atomic Shield: Excerpts from the Office Diary of Gordon E. Dean* (Chapel Hill: University of North Carolina Press, 1987), p. 19. Daniel Hirsch and William G. Mathews, "The H-Bomb: Who Really Gave Away the Secret?" *Bulletin of the Atomic Scientists*, Vol. 46 (January–February, 1990), p. 24.

27. Interview with Alice and Cyril Smith.

28. Report by John R. Murphy, February 8, 1950, KF, 65–58805, Vol. 5, Serials, 253–300.

29. Charlton C. McSwain, Report Foocase, February 15, 1950, KF, 65–58805, Vol. 7, Serials 386–405.

30. Memorandum, Re: Rudolf Ernst Peierls, January 24, 1951, KF, 65–58805, Vol. 41, Serials, 1457–1500.

31. Chadwick to Fuchs, January 24, 1946, AB1/444, PRO; Eugene Rabinowitch, "Atomic Spy Trials: Heretical Afterthoughts," *Bulletin of the Atomic*

Scientists, Vol. VII (May 1951), p. 139.

32. "Fuchs at Los Alamos," *The Spectator*, Vol. CCIII (September 18, 1959), p. 263.
33. Interview with J. Carson Mark.
34. Cited in the Washington *Star*, February 5, 1950, p. A4.
35. Conversation with Dr Mel Merritt, January 23, 1989; Lansing Lamont Papers, HST Library.
36. Brode, "Tales of Los Alamos," *LASL Community News*, July 28, 1960, p. 9.
37. K. Fuchs to G.A. McMillan, 29 November 1945, AB1/444, PRO; Richard Baker in Los Alamos *Science* (Winter/Spring 1983), p. 40; Moorehead, *The Traitors*, p. 56.
38. Moss, *Klaus Fuchs: A Biography*.
39. Smith interview; *In the Matter of J. Robert Oppenheimer* (Cambridge, Mass.: MIT Press, 1971), p. 626; Rudolph Peierls, *Bird of Passage: Recollections of a Physicist* (Princeton: Princeton University Press, 1985), p. 163; Lansing Lamont, *Day of Trinity* (New York: Atheneum, 1965), p. 18. Edward Teller to Maria Mayer, February 9, 1944, LANL Archives.
40. Smith interview; Peierls, *Bird of Passage*; Washington *Star*, February 5, 1950, p. A4; New York *Sun*, February 10, 1950; Robert Jungk, *Brighter Than a Thousand Suns: A Personal History of the Atomic Scientists* (New York: Harcourt Brace, 1956), p. 187c.
41. Deutsch statement, KF.
42. Richard P. Feynman, *"What Do You Care What Other People Think?" Further Adventures of a Curious Character* (New York: W.W. Norton, 1988), pp. 50–2.
43. Re: Foocase, April 4, 1950, KF, 65–58805, Vol. 17, Serials, 826–846.
44. Peggy Titterton interview.
45. Undated statement by Deutsch BS [65–3319], KF, 65–58805, Vol. 8, Serials, 406–475.
46. *In the Matter of J. Roberts Oppenheimer*, pp. 494–651; a piece of doggerel in Harwell read: "Fuchs looks / an ascetic theoretic," *Newsweek*, Vol. 35 (March 13, 1950), p. 34.
47. Cited in Leslie R. Groves, *Now It Can Be Told: The Story of the Manhattan Project* (New York: Da Capo Press, 1962; 1983), p. 139.
48. Chapman Pincher, *Traitors: The Anatomy of Treason* (New York: St. Martin's Press 1987), pp. 320–1n.
49. Pincher, *Too Secret Too Long*, p. 94; Charles Curran, "Fuchs at Los Alamos," *Spectator* (September 18, 1959), pp. 363-4, is disappointing; Herbert York, *The Advisors: Oppenheimer, Teller and the Superbomb* (San Francisco: W.H. Freeman, 1976), p. 37, feels that espionage played little role in the creation of the Soviet H-bomb. This thesis is convincingly supported by Daniel Hirsch and William G. Mathews, "The H-Bomb: Who Really Gave Away the Secret?" *Bulletin of the Atomic Scientists*, Vol. 46 (January–February, 1990), pp. 23–30.
50. *In the Matter of J. Robert Oppenheimer*, p. 175.
51. Bethe statement in Lansing Lamont Papers, HST Library.
52. Washington *Star*, February 5, 1950, p. A4.
53. Stewart Cockburn and David Ellyard, *Oliphant: The Life and Times of Sir Mark Oliphant* (Adelaide: Axiom Books, 1981), pp. 35–6. Lawrence Badash,

Kapitza, Rutherford and the Kremlin (New Haven: Yale University Press, 1985); David Holloway, "Entering the Nuclear Arms Race: The Soviet Decision to Build the Atomic Bomb, 1939–1945," *Social Studies of Science*, Vol. 11 (1981), p. 169. See also David Holloway, *The Soviet Union and the Arms Race* (New Haven: Yale University Press, 1983; 1986).

54. Andrew Sinclair, *The Red and the Blue: Intelligence, Treason and the Universities* (London: Weidenfeld & Nicolson, 1986) as reviewed by Robert Skidelsky, *The Sunday Times*, June 22, 1986.

55. Arnold Kramish, *Atomic Energy in the Soviet Union* (Stanford: Stanford University Press, 1959), p. 21.

56. An excellent summary may be found in David Holloway, "Science and Power in the Soviet Union," in Nicolaas A. Rupke (ed.), *Science, Politics and the Public Good: Essays in Honour of Margaret Gowing* (London: Macmillan, 1988), pp. 141–59.

57. David Holloway, "Entering the Nuclear Arms Race: The Soviet Decision to Build the Atomic Bomb, 1939–1945," pp. 159–97.

58. Cited in Holloway, *The Soviet Union and the Arms Race*, p. 20.

59. Anatoly Sonin, "How the A-Bomb Saved Soviet Physicists' Lives," *Moscow News*, weekly, No. 13, 1990. Copy lent by Richard G. Robbins.

60. The importance of the Germans in the Russian bomb is discussed in Ulrich Albrecht, "The Development of the First Atomic Bomb in the USSR." Paper for the Conference at Harvard University, Cambridge, January 8–10, 1987. Copy lent by Robert Del Tredici.

61. Oppenheimer in *In the Matter of J. Robert Oppenheimer*, p. 220.

62. Albrecht, "The Development of the First Atomic Bomb in the USSR," p. 29.

63. Bob Considine, "How Russia Stole America's Atomic Secrets" (pamphlet, 1951), KF.

64. Hyde, *The Atom Bomb Spies*, pp. 81–3, 118.

65. Eugene Rabinowitch, "Atomic Spy Trials: Heretical Afterthoughts," *Bulletin of the Atomic Scientists*, VII (May, 1951), p. 141; Philip Knightley, *The Second Oldest Profession: Spies and Spying in the Twentieth Century* (New York: W.W. Norton, 1986), pp. 165–266.

66. New York *Times*, January 29, 1988.

67. See, for example, Donald Sutherland, *The Great Betrayal: The Definitive Story of the Most Sensational Spy Case of the Century* (New York: Penguin, 1980); Peter Wright, *Spycatcher: The Candid Autobiography of a Senior Intelligence Officer* (New York: Viking, 1987); John Costello, *Mask of Treachery* (New York: Morrow, 1988); Robert Cecil, *A Divided Life: A Biography of Donald Maclean* (1988); Nigel West, *Molehunt: Searching for Soviet Spies in MI5* (1986); Philip Knightley, *The Master Spy: The Story of Kim Philby* (1988); Alan Bennett has written *Single Spies* for the National Theatre. Noel Annan gives a good review of the latest works in *The New York Review of Books*, April 13, 1989, as does Robin Winks in the *New York Times* book review, April 16, 1989.

68. Moorehead, *The Traitors,* quoted on p. xvii; *The Times*, October 10, 1960.

69. San Francisco *Chronicle*, undated copy, author's collection.

70. Albuquerque *Tribune*, August 5, 1988, August 17, 1988, September 13, 1988.

CHAPTER 7   THE BRITISH MISSION AND THE POSTWAR
             NUCLEAR CULTURE

1. Stephane Groueff, *Manhattan Project: The Untold Story of the Making of the Atomic Bomb* (New York: Bantam, 1967), p. 238.

2. Andrew J. Pierre, *Nuclear Politics: The British Experience with an Independent Strategic Force, 1939–1970* (London: Oxford University Press, 1972), pp. 52–3.

3. Richard G. Hewlett and Oscar E. Anderson, Jr., *A History of the United States Atomic Energy Commission*, Vol. I: *The New World, 1939/1946* (University Park, Penn.: Pennsylvania State University Press, 1962), p. 310.

4. Peter Malone, *The British Nuclear Deterrent* (New York: St. Martin's Press, 1984), p. 50; John Cockcroft also noted that the British were the first to say definitely that the bomb would work. In *The Hinge of Fate*, (Boston: Houghton Mifflin, 1950), Churchill said that he had "no doubt that it was the progress that we had made in Britain and the confidence of our scientists in ultimate success" that led Franklin Roosevelt to inaugurate the American nuclear effort (p. 381).

5. Sir John Cockcroft, review of Gowing in *Disarmament and Arms Control* III (Spring 1965), 89; Cockcroft, quoted in Martin Gilbert, *Road to Victory*, 487n.

6. French interview.

7. Moon interview.

8. Peierls interview.

9. Smith interview; Porton interview; conversation with Norris Bradbury, spring 1985; N.E. Bradbury, Certificate for William G. Penney, 9 May 1946, LANL Archives; Robert Oppenheimer, "Niels Bohr and Atomic Weapons," *New York Review of Books*, Vol. III (December 17, 1964), p. 6; Oppenheimer statement in the Santa Fe *New Mexican*, August 17, 1945; copy reproduced in Fern Lyon and Jacob Evans, *Los Alamos: The First Forty Years* (Los Alamos Historical Society, 1984), p. 42.

10. *In the Matter of J. Robert Oppenheimer* (Cambridge, Mass.: MIT Press, 1971), p. 177; Leslie R. Groves, *Now It Can Be Told: The Story of the Manhattan Project* (New York: Da Capo, 1983 [1967]), p. 408.

11. Bethe letter cited in Dennis C. Eakley, "The British Mission," *Los Alamos Science* (Winter/Spring 1983), p. 187.

12. Manley interview.

13. French to Author, August 18, 1989.

14. Penney interview.

15. Margaret Gowing, *Britain and Atomic Energy, 1939–1945*, (London: Macmillan, 1964), p. 267; John Baylis, *Anglo-American Defense Relations, 1939–1984: The Special Relationship* (New York: St. Martin's Press, 1984), pp. 31, 176. See also Ronald W. Clark, *The Birth of the Bomb: The Untold Story of Britain's Part in the Weapon That Changed the World* (London: Phoenix House, 1960).

16. The phrase is from a guest editorial, January 22, 1949, in the Rochester *Times Union*, by Robert E. Marshak. FBI papers. See also Paul Boyer, *By the Bomb's Early Light*; cf. V.P. Keay to H.B. Fletcher, February 9, 1950, FBI, 65–58805, Vol. 4, Serials, 172–252.

17. Quoted in *Physics Today* (September, 1989), p. 82.
18. Rotblat interview.
19. *Guardian*, August 1, 1986; J. Rotblat, *History of the Pugwash Conferences* (London: Taylor & Francis, 1962).
20. *Physics Today* (September, 1989), p. 82; William Gutteridge, Andrew Haines, and Anthony de Rouck, *Pugwash: Shaping Its Future* (newsletter, 1989).
21. London *Times*, August 14, 1945.
22. Gowing, *Independence and Deterrence*, Vol. II: *Policy Execution,* pp. 120–1.
23. Hans Bethe and Frederick Seitz, "How Close Is the Danger?" *One World or None* (New York: McGraw-Hill, 1946), pp. 42–6.
24. London *Times*, September 18, 1945.
25. Bohr, as quoted by David Holloway, *The Soviet Union and the Arms Race* (New Haven: Yale University Press, 1983, 1986), p. 23.
26. Chase S. Osborn to Truman, December 11, 1945. Official Files, 692 An "Atomic Bomb," HST Library.
27. Albuquerque *Journal* April 2, 1946.
28. Albuquerque *Journal* March 24, 1946.
29. Cf. Floyd Gottfredson, "Mickey Mouse on Sky Island" (1936–7), in *Mickey Mouse* (New York: Abbeville Press, 1978); Felix Marley, "Red Herrings Return to Roost," *Human Events: A Weekly Analysis for the American Citizen,* Vol. VII (February 8, 1950), n.p.; Bob Considine, "How Russia Stole America's Atomic Secrets," pamphlet, 1951.
30. M.F. Perutz, "That Was the War: Enemy Alien," *New Yorker* (August 12, 1985), p. 38.
31. Teletype, Foocase, April 29, 1950, KF, 65–58805, Vol. 32, Serials, 1187–1220.
32. Wright, *Spycatcher*, p. 237.
33. J. Edgar Hoover, "Memorandum for Mr. Tolson [and] Mr Ladd," March 2, 1950, KF, 65–58805, Vol. 12, Serials, 586–641.
34. Albuquerque *Tribune*, September 14, 1989.
35. Alice Kimball Smith, "Los Alamos: Focus of an Age," *Bulletin of the Atomic Scientists* (June 1970), p. 18; Boyer, *By the Bomb's Early Light, passim.*
36. *Psychosocial Aspects of Nuclear Developments: A Report of the Task Force on Psychosocial Aspects of Nuclear Developments of the American Psychiatric Association* (Washington: American Psychiatric Association, 1982), p. v.
37. Rudolf Peierls, "Forty Years into the Atomic Age," in Nicolaas A. Rupke (ed.), *Science, Politics and the Public Good: Essays in Honour of Margaret Gowing* (London: Macmillan, 1988), pp. 160–8.
38. Albuquerque *Journal*, July 16, 1990, p. A5.
39. McGeorge Bundy, *Danger and Survival: Choices About the Bomb in the First Fifty Years* (New York: Random House, 1988), pp. 597, 600–1.
40. R.P. Turco, O.B. Toon, T.P. Ackerman, J.B. Pollack and Carl Sagan [TTAPS], "Global Atmospheric Consequences of Nuclear War," *Science*, Vol. 222 (December 23, 1983), pp. 1283–1291.
41. Sagan in Charles W. Kegley, Jr., and Eugene R. Wittkopf (eds), *The Nuclear Reader: Strategy, Weapons, War* (New York: St. Martin's Press, 1985), p. 329. See also the essays in Catherine McArdle Kelleher *et al.* (eds), *Nuclear*

Deterrence: New Risks, New Opportunities (Washington: Pergamon-Brassey's, 1986).

42. Helen Caldicott, *Missile Envy: The Arms Race and Nuclear War* (New York: William Morrow, 1984); James Thompson, *Psychological Aspects of Nuclear War* (New York: The British Psychological Society and John Wiley 1985).

43. Rita R. Rogers, "On Emotional Responses to Nuclear Issues and Terrorism," in *Psycholocial Aspects*, pp. 11–23.

44. Robert Jay Lifton, "Imagining the Real: Beyond the Nuclear 'End,'" in Lester Grinspoon (ed.), *The Long Darkness: Psychological and Moral, Perspectives on Nuclear Winter* (New Haven: Yale University Press, 1986), pp. 81ff.

45. Lifton, "Imagining the Real," p. 91.

46. Tuck to Alice Kimball Smith, no date but *c.* January, 1970, Tuck MSS, LANL Archives.

47. Rotblat in *The Times Literary Supplement* (*Education Supplement*) (December 20, 1981), p. 18.

48. Lorna Arnold, *A Very Special Relationship: British Atomic Weapon Trials in Australia* (London: HMSO, 1987), p. 247.

49. Bethe and Oppenheimer quoted in *Newsweek*, July 19, 1965, pp. 50–3; interview with Ellen Bradbury.

50. Oppenheimer in *Newsweek*, July 19, 1945, p. 51; Mark, quoted in Ferenc Morton Szasz, *The Day the Sun Rose Twice: The Story of the Trinity Site Nuclear Explosion, July 16, 1945* (Albuquerque: University of New Mexico Press, 1984), p. 126; Hans Bethe, "Chop Down the Nuclear Arsenals," *Bulletin of the Atomic Scientists*, Vol. 45 (March 1989), p. 12.

51. *Guardian*, August 2, 1986; cf. Rudolf Peierls, "Reflections of a British Participant," *Bulletin of the Atomic Scientists*, Vol. 41 (August 1985), pp. 27–9, and "Britain in the Atomic Age," in Richard S. Lewis and Jan Wilson (eds), *Alamogordo Plus Twenty-five Years* (New York: Viking, 1971), pp. 95–6.

Appendix I
The Postwar Careers of the British Mission

NIELS BOHR

Bohr's career has already been sketched in Chapter 4. When he died in 1962, the world of science lost one of its greatest minds. The columnists were lavish in their praise, for Bohr was one individual of whom nobody seemed to have an ill word. They might disagree with his ideas, but they always respected the person behind them. His essays and speeches are available in a three-volume set, *The Philosophical Writings of Niels Bohr*. His son, Aage, continues to direct his father's institute in Copenhagen. Aage Bohr won the Nobel Prize in 1975.

EGON BRETSCHER

Egon Bretscher returned to England where he worked at Harwell for many years. He was Head of the Nuclear Physics Division from 1948 until his retirement in 1966. He also lectured at Cambridge, and authored numerous papers in chemistry, atomic physics, solid state physics, and nuclear physics. He died in 1973. His manuscripts are deposited in the Churchill College Archives Centre, Cambridge, England. His wife, Hanni, still lives in Cambridge.

JAMES CHADWICK

James Chadwick returned to Liverpool exhausted from his war efforts. In 1945, he was knighted for his services to Britain. On his return to Liverpool, he tried to develop a nuclear physics school on a large scale and succeeded, amidst severe economic shortages, in obtaining both a new building and a cyclotron for his program. In 1948, he moved to Cambridge as Master of Gonville and Caius College and devoted the rest of his career to academic administration. He retired from Cambridge in 1958 and in 1970 was made a Companion of Honour. While he published numerous studies on radioactivity, he unfortunately did not live to complete his autobiography. He died in 1974. His papers are at Churchill College, Cambridge.

LORD CHERWELL

By 1939, Frederick Alexander Lindemann had established his reputation as one of Britain's foremost scientists. A man of brilliant and eccentric behavior, he became an instant legend in 1914 when he devised the safest way of pulling an airplane out of a spin. "When flying ordinarily," a London *Times* reporter noted on his death, "it should be said he was not a very good pilot."

For years the director of the Clarendon Laboratory at Oxford, Lindemann was described as "Old Man River" because of his silences. In 1939, when Churchill assumed the post of Prime Minister, he appointed Lindemann as his personal assistant (later as Paymaster-General). In 1941, Lindemann was created Baron Cherwell. Cherwell advised Churchill all through the conflict on a wide range of scientific matters. Cherwell advised Churchill again from 1951–3, and in 1954 was appointed a member of the United Kingdom Atomic Energy Authority, the same year he was made a Companion of Honour. A staunch advocate of above-ground weapons testing, Cherwell dismissed the concept of harmful fall-out as "unmitigated nonsense." He died in 1957.

BORIS DAVISON

Born in 1908 in Russia of a British father, Davison joined the British Mission in Los Alamos in October, 1945. After the war, he returned to the British Isles to work at the Atomic Energy Research Establishment at Harwell. The Fuchs case affected him considerably. Because his parents were still behind the Iron Curtain, and might presumably be used in a blackmail attempt, Davison had to resign from Harwell. A year's leave of absence was spent at the University of Birmingham. In 1954, he returned to North America to become Associate Professor of Physics and Mathematics at the University of Toronto Computation Center. He died in 1960. In 1979, Atomic Energy of Canada Limited published the *Collected Papers of Boris Davison*.

A.P. FRENCH

French returned to England to work at Harwell, where he stayed for two years. While there he wrote up part of his Los Alamos research as a Cambridge PhD dissertation. In 1948, he began teaching and doing nuclear research at the Cavendish Laboratory in Cambridge and by the end of the first term had received his doctorate.

Seven years later, French, with his American-born wife, returned to the USA to join the physics department at the University of South Carolina. In 1962, he accepted a visiting appointment at MIT and in 1964 joined the permanent faculty there as Professor of Physics. His major publications include *Newtonian Mechanics* (1970) and *An Introduction to Quantum Physics* (1979), in addition to two edited works: *Niels Bohr: A Centenary Volume* and *Einstein: A Centenary Volume*. Because of his work in physics education, French has been widely acknowledged as the "foremost teacher of physics" for this generation. He lives in Cambridge, Massachusetts.

OTTO FRISCH

Otto Frisch returned to Great Britain after the war where he became leader of the Division of Experimental Physics at Harwell. While the laboratory was being built, he did valuable calculations on the fluctuations of a nuclear pile, and their effect on oscillator tests. In 1947, he assumed a chair in Cambridge, serving as Jacksonian Professor of Natural Philosophy for the remainder of his career. He never worked on weapons matters again. Instead, Frisch devoted his efforts to instrumentation and to education in science.

Among his books are *The Nuclear Handbook* (1955) and the popular *The Nature of Matter* (1973). After much urging, his friends convinced him to write down his favorite stories. In 1979, his fascinating autobiography *What Little I Remember* broke into print, one year before his death.

KLAUS FUCHS

In 1959, Fuchs was freed from jail and immediately flew to East Germany. Soon after his arrival he married Greta Keilson, a German communist and longtime friend. Fuchs then took a position in the East German Central Institute for Nuclear Physics at Rossendorf near Dresden, and he headed the Center from 1974 until his retirement in 1979. Fuchs was well respected in East Germany. He owned two cars and lived for twenty years in a villa overlooking the Elbe, as befitted a member of the East German Communist Party Central Committee. Upon his death in 1988 the official East German news agency praised him as a man who had "devoted his whole life to the struggle of the party of the working class." The official eulogy released to the western press praised his scientific achievements in the field of theoretical physics, his actions for socialism and world peace, his friendship with the Soviet Union, and his twenty years of development of the power industry. It did not mention his espionage.

JAMES HUGHES

After his Los Alamos experience, James Hughes returned to the University of Liverpool to complete his PhD. After a post-doctoral fellowship, he taught as a member of the Department of Physics at Liverpool from 1949 to 1951, during which time he published several scholarly articles. He subsequently moved to Edinburgh, where he served as a lecturer in physics for about five years. The world of industry then beckoned and Hughes joined it for over a decade and a half. He eventually grew disillusioned with corporate life, however, and returned to teaching, this time in the state system in Liverpool. He retired from secondary teaching in 1979 and died a few years later.

DERRIK JOSEPH LITTLER

After the war, Littler joined the British Atomic Energy Research Establishment at Harwell, where he held many positions. He became a specialist in low energy reactors, served as British Physics Secretary for the 1955 "Atoms for Peace" conference in Geneva and headed up a "reactor school" at Harwell that was designed to teach engineers both nuclear and reactor physics. In 1959, he joined the Central Electricity Generating Board (CEGB). From 1970–81, he served as chief physicist in their research department, and also edited, for a time, the *Journal of Nuclear Energy*. From 1981–3, he was principal of the CEGB management staff college. Littler retired in 1983 and currently lives in Middlesex.

J. CARSON MARK

Born in Canada in 1913, Mark became an American citizen after the war. When Bethe left Los Alamos, George Placzek became the head of the Theoretical Division, but he soon passed that assignment on to Mark. From 1947 until his retirement, Mark headed the Theoretical Division at Los Alamos. He presided over this section of Los Alamos at the time of miniaturization of weapons. He wrote several technical papers and played a major role in modern weapons research. In 1958, he served as a U.S. Delegate to a Conference of experts on the means of Detection of Nuclear Explosions. He also served as a member of the advisory committee for Reactor Safeguards of the U.S. Nuclear Regulatory Commission and as a member of the Scientific Advisory Board for the U.S. Air Force. He and his wife Kathleen still live in Los Alamos.

WILLIAM GREGORY MARLEY

Marley made his chief contribution to the Manhattan Project through his work in high-speed photography. But when he arrived at Harwell in 1946, as one of the first three scientists to land on the disused airfield, he turned to another area: health physics. In 1948, he formed the Health Physics Division of Harwell and later became head of the Radiological Protection Division of the United Kingdom Atomic Energy Authority Health and Safety Branch. Eventually, he became Harwell's chief radiological expert. Internationally, he contributed to establishing radiological protection standards throughout the world. In 1966, he received the Elda E. Anderson Award from the American Health Physics Society. He died in 1980.

DONALD G. MARSHALL

Marshall returned to the University of Birmingham for a PhD and taught for a few years as a lecturer in the Physics Department. But he found academic life a bit slow after Los Alamos and soon joined the Dunlop Corporation, where he spent his entire career. He rose to Section Manager of the Research Laboratory, then Manager of the

Laboratory, Technical Manager of Tyres, UK., Research Manager of the Central Division (second in command to the director) and, finally, to Director of Central Research. He retired from the latter position in 1980. Marshall currently lives in the West Midlands.

PHILIP B. MOON

Moon returned to England and in 1950 was appointed Poynting Professor of Physics at the University of Birmingham, where he succeeded Professor Marcus Oliphant, who returned to Australia. Moon taught his entire career at Birmingham, during which time he received numerous research grants. Immediately after the war, he became a major spokesman for atomic science in the British Isles, and served for years as the secretary of the Atomic Scientists Association. He also spoke out strongly on the issue of free exchange of scientific information. In 1974, the Department of Physics of the University of Birmingham held a conference on "Scientific Aspects of High Speed Rotation," a field in which he had done important work and had exploited for nuclear collision measurements. In his honor, papers from the symposium were later published. P.B. Moon wrote numerous scientific studies, including *Artificial Radioactivity* (1949) and *Ernest Rutherford and the Atom* (1979). He currently lives in Shropshire.

RUDOLF PEIERLS

Peierls returned to England to become Professor of Mathematical Physics at Birmingham. In 1963, he became Wykeham Professor at the University of Oxford and a Fellow of New College. After retirement in 1974 he taught at the University of Washington in Seattle for three years. He has also taught at UCLA and Cal Tech. Peierls received numerous honors, including a knighthood in 1968. An admirer termed him "one of the last great physicists who refused to specialize in a narrow field." In 1980, B.H. Bransden, Professor of Physics at the University of Durham, called Peierls "the doyen of theoretical physics in the United Kingdom." When he retired from Oxford in 1974, the occasion was celebrated by a symposium later published as *Rudolf Peierls and Theoretical Physics*.

Peierls was the author of numerous volumes, but his most famous were *The Laws of Nature* (1955), an attempt to explain physics in lay language, and *Surprises in Theoretical Physics* (1979). His memoir, *Bird of Passage: Recollections of a Physicist* is delightful. He resides in Oxford.

WILLIAM GEORGE PENNEY

The highlights of Penney's career have already been sketched in Chapter 4. He directed the Atomic Weapons Research Establishment, Aldermaston, from 1953 to 1959 when he became Authority Member for Scientific Research at Harwell. He became Deputy Chairman of the UK Atomic Energy Authority in 1961 and Chairman from 1964 to 1967. He left that spot to become rector of Imperial College, London, in 1967. He retired in 1972.

Penney has received numerous honors. He was awarded honorary degrees from Melbourne University, Bath University of Technology, and the Universities of Durham and Oxford. In 1967, he was created a life peer. Two years later the Queen awarded him the Order of Merit. An acknowledged world authority in his field, he lived near Oxford until his death on March 4, 1991.

GEORGE PLACZEK

Born in Czechoslovakia in 1905, Placzek eventually took out American citizenship and continued his work in neutron diffusion theory in the States. In December 1945, he succeeded Bethe as leader of the Theoretical Division at Los Alamos. He had hoped to remain in Los Alamos but his heart could not stand the altitude, and he left in July 1946. He was even reluctant to return to the Hill on a consulting basis. From 1946 to 1948, he worked at the Research Laboratory of General Electric in Schenectady, New York. Later he moved to the Institute for Advanced Study in Princeton. During the hunt for Fuchs's American contact, the FBI investigated Placzek thoroughly. His hatred of communists (he called them "animals and liars") plus strong support from I.I. Rabi helped clear him of suspicion. He wrote numerous scientific and technical essays on neutron diffusion. Placzek died in 1955.

MICHAEL J. POOLE

Poole returned from Los Alamos to Britain in October, 1945, and later, in 1947–9, wrote up his Los Alamos work for a PhD thesis. Poole remained at Harwell for the rest of his professional life. After retirement in 1985, he became a professor of Physics at Oxford for three years. He retired from Oxford in 1987, and currently lives in Abingdon on Thames.

JOSEPH ROTBLAT

Rotblat's postwar career has been delineated in Chapter 4. The success of Pugwash has given him considerable publicity and he has begun to write his memoirs. Rotblat should also be noted as the author of several books, among which are: *Pugwash, the First Ten Years* (1967); *Atoms and the Universe* (1956; 1973); *Science and World Affairs* (1962); *Nuclear Reactors: To Breed or Not To Breed* (1977); and *Scientists, the Arms Race, and Disarmament* (1982).
He lives in London.

HAROLD SHEARD

Born in 1909, Sheard earned an MSc in physics from the University of London.

After Los Alamos, he took a job at Harwell in the Reactor Physics Division from 1952 to 1955, at which time he moved to Canada for three years as the UKAEA representative there. After a brief stint as Technical Manager of Windscale, he became the Chief Physicist at the Radiochemical Centre, Amersham, Bucks., from 1959 to 1964; he served as their Deputy Director from 1964 until his retirement in 1969. He died in 1987.

TONY HILTON ROYLE SKYRME

Skyrme returned to England in February of 1946 for two years' service as a Research Fellow at the University of Birmingham. In 1948, he returned to the US for a year of study under Victor Weisskopf at MIT. From there he moved to become a Research Fellow at the Institute for Advanced Study in Princeton.

From 1950 to 1961, Skyrme headed the nuclear physics group, Theoretical Physics Division, at Harwell. After two years as Senior Lecturer at the Department of Mathematics, University of Malaya, Pantai Valley, Kuala Lumpur, Malaysia, Skyrme returned to become Professor of Mathematical Physics, and later Professor of Applied Mathematics, at the University of Birmingham. During his research he hypothesized a sub-atomic particle later named the "skyrmion." In 1985, the Royal Society awarded him the Hughes Medal for his contributions to science. Author of over thirty technical papers, few of Skyrme's works have broken into the public realm. He died in 1987.

SIR GEOFFREY I. TAYLOR

For over fifty years, G.I. Taylor made steady contributions to the field of British science. His reputation extended to applied mathematics, classical physics and engineering. During World War II, he worked on numerous defense problems; at the close of hostilities, he began to investigate whatever issue caught his fancy. In 1952, he resigned from the Royal Society's Yarrow Research Professorship, which he had held since 1923, but the Society continued to support his experiments. These new interests ranged from analysis of the swimming style of spermatozoa to the movement of gas bubbles through water to the problems of fluid dynamics. In 1969, he was appointed to the Order of Merit. Over the course of his life, Taylor wrote perhaps two hundred scientific papers, which Cambridge University Press later published in four large volumes. A biographer termed Taylor "one of the most notable scientists of this century." He died in 1975.

ERNEST W. TITTERTON

Titterton's career has been sketched in Chapter 4. In 1946, the Tittertons returned to England, and they remained at Harwell for four years. In 1950, he emigrated to Australia to become head of the Nuclear Physics Department at the Australian National University in Canberra, where he later became Dean of the ANU Research

School of Physical Sciences. He also became a Fellow of the Royal Australian Academy of Science. He received many honors, including a knighthood. Author of over 200 scientific papers, his works include: *Facing the Atomic Future* (1956) and *Uranium, Energy Source of the Future* (1974).

In September, 1987, Sir Ernest was involved in a severe automobile accident that left him a quadriplegic. After several months in a Sydney Hospital, he moved to the Canberra area to be near his family. Utilizing a voice-activated tape recorder, he set to work on his autobiography, but died in 1990.

JAMES TUCK

After the war James Tuck returned to Oxford, where he served as Supervisor for the Department of Advanced Studies, Clarendon Laboratory, from 1946 to 1949. King George VI awarded him an OBE for his war work. Afterwards, Tuck moved to the Institute for Nuclear Studies at the University of Chicago for a year. When he became a US citizen, he and his family returned to Los Alamos. "The place had changed," he noted; "it was less glamorous but the old charm was still operative." His wife Elsie agreed. She often said that Los Alamos was the only American town in which she would live. Their home was known as "Little England" on the Hill.

Tuck spent the rest of his career as Associate Division Leader of the Physics Division at Los Alamos, where he served from 1950 to 1973. Over the years he became a Los Alamos legend, both for the brilliance of his discoveries as well as the eccentricity of his behavior. On December 15, 1980, James Tuck died after a long illness; his wife Elsie followed six months later.

Appendix II
The Frisch-Peierls
Memorandum (March 1940)

[This is the first time the entire Frisch-Peierls Memorandum has been published as a unit. Part I comes from Gowing, *Britain and Atomic Energy, 1939–1945*, pp. 389–93; Part II from Clark, *Tizard*, pp. 214–19 (see Bibliography).]

On the Construction of a 'Super-bomb'; based on a
Nuclear Chain Reaction in Uranium

The possible construction of 'super-bombs' based on a nuclear chain reaction in uranium has been discussed a great deal and arguments have been brought forward which seemed to exclude this possibility. We wish here to point out and discuss a possibility which seems to have been overlooked in these earlier discussions.

Uranium consists essentially of two isotopes, $^{238}U(99.3\%)$ and $^{235}U(0.7\%)$. If a uranium nucleus is hit by a neutron, three processes are possible: (1) scattering, whereby the neutron changes direction and, if its energy is above 0.1 MeV, loses energy; (2) capture, when the neutron is taken up by the nucleus; and (3) fission, i.e. the nucleus breaks up into two nuclei of comparable size, with the liberation of an energy of about 200 MeV.

The possibility of a chain reaction is given by the fact that neutrons are emitted in the fission and that the number of these neutrons per fission is greater than 1. The most probable value for this figure seems to be 2.3, from two independent determinations.

However, it has been shown that even in a large block of ordinary uranium no chain reaction would take place since too many neutrons would be slowed down by inelastic scattering into the energy region where they are strongly absorbed by ^{238}U.

Several people have tried to make chain reaction possible by mixing the uranium with water, which reduces the energy of the neutrons still further and thereby increases their efficiency again. It seems fairly certain, however, that even then it is impossible to sustain a chain reaction.

In any case, no arrangement containing hydrogen and based on the action of slow neutrons could act as an effective super-bomb, because the reaction would be too slow. The time required to slow down a neutron is about 10^{-5} sec and the average time lost before a neutron hits a uranium nucleus is even 10^{-4} sec. In the reaction, the number of neutrons would increase exponentially, like $e^{\,t}/\tau$ where τ would be at least 10^{-4} sec. When the temperature reaches several thousand degrees the container of the bomb will break and within 10^{-4} sec the uranium would have expanded sufficiently to let the neutrons escape and so to stop the reaction. The energy liberated would, therefore, be only a few times the energy required to break the container, i.e., of the same order of magnitude as with ordinary high explosives.

Bohr has put forward strong arguments for the suggestion that the fission observed with slow neutrons is to be ascribed to the rare isotope ^{235}U, and that this isotope has, on the whole, a much greater fission probability than the common isotope ^{238}U. Effective methods for the separation of isotopes have been developed recently, of which the method of thermal diffusion is simple enough to permit separation on a fairly large scale.

This permits, in principle, the use of nearly pure ^{235}U in such a bomb, a possibility which apparently has not so far been seriously considered. We have discussed this possibility and come to the conclusion that a moderate amount of ^{235}U would indeed constitute an extremely efficient explosive.

The behaviour of ^{235}U under bombardment with fast neutrons is not known experimentally, but from rather simple theoretical arguments it can be concluded that almost every collision produces fission and that neutrons of any energy are effective. Therefore it is not necessary to add hydrogen, and the reaction, depending on the action of fast neutrons, develops with very great rapidity so that a considerable part of the total energy is liberated before the reaction gets stopped on account of the expansion of the material.

The critical radius r_0 – i.e. the radius of a sphere in which the surplus of neutrons created by the fission is just equal to the loss of neutrons by escape through the surface – is, for a material with a given composition, in a fixed ratio to the mean free path of the neutrons, and this in turn is inversely proportional to the density. It therefore pays to bring the material into the densest possible form, i.e., the metallic state, probably sintered or hammered. If we assume, for ^{235}U, no appreciable scattering, and 2.3 neutrons emitted per fission, then the critical radius is found to be 0.8 times the mean free path. In the metallic state (density 15), and assuming a fission cross-section of 10^{-23} cm^2, the mean free path would be 2.6 cm and r_0 would be 2.1 cm, corresponding to a mass of 600 grams. A sphere of metallic ^{235}U of a radius greater than r_0 would be explosive, and one might think of about 1 kg as a suitable size for the bomb.

The speed of the reaction is easy to estimate. The neutrons emitted in the fission have velocities of about 10^9 cm/sec and they have to travel 2.6 cm before hitting a uranium nucleus. For a sphere well above the critical size the loss through neutron escape would be small, so we may assume that each neutron, after a life of 2.6×10^{-9} sec, produces fission, giving birth to two neutrons. In the expression ε^{τ}/τ for the increase of neutron density with time, it would be about 4×10^{-9} sec, very much shorter than in the case of a chain reaction depending on slow neutrons.

If the reaction proceeds until most of the uranium is used up, temperatures of the order of 10^{10} degrees and pressure of about 10^{13} atmospheres are produced. It is difficult to predict accurately the behaviour of matter under these extreme conditions, and the mathematical difficulties of the problem are considerable. By a rough calculation we get the following expression for the energy liberated before the mass expands so much that the reaction is interrupted:

$$E = 0.2M(r^2/\tau^2)\ (\sqrt{(r/r_0)} - 1) \tag{1}$$

(M, total mass of uranium; r, radius of sphere; r_0 critical radius; τ, time required for neutron density to multiply by a factor e). For a sphere of diameter 4.2 cm ($r = 2.1$ cm), M = 4,700 grams, $\tau = 4 \times 10^{-9}$ sec, we find E = 4×10^{22} ergs, which is about one-tenth of the total fission energy. For a radius of about 8 cm (M = 32 kg) the whole fission energy is liberated, according to formula (1). For small radii the efficiency falls off even faster than indicated by formula (1) because τ goes up as r approaches r_0. The energy liberated by a 5 kg bomb would be equivalent to that of several thousand tons of dynamite, while that of a 1 kg bomb, though about 500 times less, would still be formidable.

It is necessary that such a sphere should be made in two (or more) parts which are brought together first when the explosion is wanted. Once assembled, the bomb would explode within a second or less, since one neutron is sufficient to start the reaction and there are several neutrons passing through the bomb in every second, from the cosmic radiation. (Neutrons originating from the action of uranium alpha rays on light-element impurities would be negligible provided the uranium is reasonably pure.) A sphere with a radius of less than about 3 cm could be made up in two hemispheres, which are pulled together by springs and kept separated by a suitable structure which is removed at the desired moment. A larger sphere would have to be composed of more than two parts, if the parts, taken separately, are to be stable.

It is important that the assembling of the parts should be done as rapidly as possible, in order to minimise the chance of a reaction getting started at a moment when the critical conditions have only just been reached. If this happened, the reaction rate would be much slower and the energy liberation would be considerably reduced; it would, however, always be sufficient to destroy the bomb.

It may be well to emphasise that a sphere only slightly below the critical size is entirely safe and harmless. By experimenting with spheres of gradually increasing size and measuring the number of neutrons emerging from them under a known neutron bombardment, one could accurately determine the critical size, without any danger of a premature explosion.

For the separation of the ^{235}U, the method of thermal diffusion, developed by Clusius and others, seems to be the only one which can cope with the large amounts required. A gaseous uranium compound, for example uranium hexafluoride, is placed between two vertical surfaces which are kept at a different temperature. The light isotope tends to get more concentrated near the hot surface, where it is carried upwards by the convection current. Exchange with the current moving downwards along the cold surface produces a fractionating effect, and after some time a state of equilibrium is reached when the gas near the upper end contains markedly more of the light isotope than near the lower end.

For example, a system of two concentric tubes, of 2 mm separation and 3 cm diameter, 150 cm long, would produce a difference of about 40% in the concentration of the rare isotope between its ends, and about 1 gram per day could be drawn from the upper end without unduly upsetting the equilibrium.

In order to produce large amounts of highly concentrated ^{235}U, a great number of these separating units will have to be used, being arranged in parallel as well as in series. For a daily production of 100 grams of ^{235}U of 90% purity, we estimate that about 100,000 of these tubes would be required. This seems a large number, but it would undoubtedly be possible to design some kind of a system which would have the same effective area in a more compact and less expensive form.

In addition to the destructive effect of the explosion itself, the whole material of the bomb would be transformed into a highly radioactive state. The energy radiated by these active substances will amount to about 20% of the energy liberated in the explosion, and the radiations would be fatal to living beings even a long time after the explosion.

The fission of uranium results in the formation of a great number of active bodies with periods between, roughly speaking, a second and a year. The resulting radiation is found to decay in such a way that the intensity is about inversely proportional to the time. Even one day after the explosion the radiation will correspond to a power expenditure of the order of 1,000 kW, or to the radiation of a hundred tons of radium.

Any estimates of the effects of this radiation on human beings must be rather uncertain because it is difficult to tell what will happen to the radioactive material after the explosion. Most of it will probably be blown into the air and carried away by the wind. This cloud of radioactive material will kill everybody within a strip estimated to be several miles long. If it rained the danger would be even worse because active material would be carried down to the ground and stick to it, and persons entering the contaminated area would be subjected to dangerous radiation even after days. If 1% of the active material sticks to the debris in the vicinity of the explosion and if the debris is spread over an area of, say, a square mile, any person entering this area would be in serious danger, even several days after the explosion.

In these estimates, the lethal dose of penetrating radiation was assumed to be 1,000 Roentgen; consultation of a medical specialist on X-ray treatment and perhaps further biological research may enable one to fix the danger limit more accurately. The main source of uncertainty is our lack of knowledge as to the behaviour of materials in such a super-explosion, and an expert on high explosives may be able to clarify some of these problems.

Effective protection is hardly possible. Houses would offer protection only at the margins of the danger zone. Deep cellars or tunnels may be comparatively safe from the effects of radiation, provided air can be supplied from an uncontaminated area (some of the active substances would be noble gases which are not stopped by ordinary filters).

The irradiation is not felt until hours later when it may be too late. Therefore it would be very important to have an organisation which determines the exact extent of the danger area, by means of ionisation measurements, so that people can be warned from entering it.

O.R. Frisch
R. Peierls

The University, Birmingham

Part II

The attached detailed report concerns the possibility of constructing a 'super-bomb' which utilises the energy stored in atomic nuclei as a source of energy. The energy liberated in the explosion of such a super-bomb is about the same as that produced by the explosion of 1,000 tons of dynamite. This energy is liberated in a small volume, in which it will, for an instant, produce a temperature comparable to that in the interior of the sun. The blast from such an explosion would destroy life in a wide area. The size of this area is difficult to estimate, but it will probably cover the centre of a big city.

In addition, some part of the energy set free by the bomb goes to produce radioactive substances, and these will emit very powerful and dangerous radiations. The effect of these radiations is greatest immediately after the explosion, but it decays only gradually and even for days after the explosion any person entering the affected area will be killed.

Some of this radioactivity will be carried along with the wind and will spread the contamination; several miles downwind this may kill people.

In order to produce such a bomb it is necessary to treat a few cwt. of uranium by a process which will separate from the uranium its light isotope (U_{235}) of which it contains about 0.7%. Methods for the separation of isotopes have recently been developed. They are slow and they have not until now been applied to uranium, whose chemical properties give rise to technical difficulties. But these difficulties are by no means insuperable. We have not sufficient experience with large-scale chemical plant to give a reliable estimate of the cost, but it is certainly not prohibitive.

It is a property of these super-bombs that there exists a 'critical size' of about one pound. A quantity of the separated uranium isotope that exceeds the critical amount is explosive; yet a quantity less than the critical amount is absolutely safe. The bomb would therefore be manufactured in two (or more) parts, each being less than the critical size, and in transport all danger of a premature explosion would be avoided if these parts were kept at a distance of a few inches from each other. The bomb would be provided with a mechanism that brings the two parts together when the bomb is intended to go off. Once the parts are joined to form a block which exceeds the critical amount, the effect of the penetrating radiation always present in the atmosphere will initiate the explosion within a second or so.

The mechanism which brings the parts of the bomb together must be arranged to work fairly rapidly because of the possibility of the bomb exploding when the critical conditions have just only been reached. In this case the explosion will be far less powerful. It is never possible to exclude this altogether, but one can easily ensure that only, say, one bomb out of 100 will fall in this way, and since in any case the explosion is strong enough to destroy the bomb itself, this point is not serious.

We do not feel competent to discuss the strategic value of such a bomb, but the following conclusions seem certain:

1. As a weapon, the super-bomb would be practically irresistible. There is no material or structure that could be expected to resist the force of the explosion. If one thinks of using the bomb for breaking through a line of fortifications, it should be kept in mind that the radioactive radiations will prevent anyone from approaching the affected territory for several days; they will equally prevent defenders from reoccupying the affected positions. The advantage would lie with the side which can determine most accurately just when it is safe to re-enter the area; this is likely to be the aggressor, who knows the location of the bomb in advance.

2. Owing to the spreading of radioactive substances with the wind, the bomb could probably not be used without killing large numbers of civilians, and this may make it unsuitable as a weapon for use by this country. (Use as a depth charge near a naval base suggests itself, but even there it is likely that it would cause great loss of civilian life by flooding and by the radioactive radiations.)

3. We have no information that the same idea has also occurred to other scientists but since all the theoretical data bearing on this problem are published, it

is quite conceivable that Germany is, in fact, developing this weapon. Whether this is the case is difficult to find out, since the plant for the separation of isotopes need not be of such a size as to attract attention. Information that could be helpful in this respect would be data about the exploitation of the uranium mines under German control (mainly in Czechoslovakia) and about any recent German purchases of uranium abroad. It is likely that the plant would be controlled by Dr K. Clusius (Professor of Physical Chemistry in Munich University), the inventor of the best method for separating isotopes, and therefore information as to his whereabouts and status might also give an important clue.

At the same time it is quite possible that nobody in Germany has yet realised that the separation of the uranium isotopes would make the construction of a superbomb possible. Hence it is of extreme importance to keep this report secret since any rumour about the connection between uranium separation and a super-bomb may set a German scientist thinking along the right lines.

4. If one works on the assumption that Germany is, or will be, in the possession of this weapon, it must be realised that no shelters are available that would be effective and could be used on a large scale. The most effective reply would be a counter-threat with a similar bomb. Therefore it seems to us important to start production as soon and as rapidly as possible, even if it is not intended to use the bomb as a means of attack. Since the separation of the necessary amount of uranium is, in the most favourable circumstances, a matter of several months, it would obviously be too late to start production when such a bomb is known to be in the hands of Germany, and the matter seems, therefore, very urgent.

5. As a measure of precaution, it is important to have detection squads available in order to deal with the radioactive effects of such a bomb. Their task would be to approach the danger zone with measuring instruments, to determine the extent and probable duration of the danger and to prevent people from entering the danger zone. This is vital since the radiations kill instantly only in very strong doses whereas weaker doses produce delayed effects and hence near the edges of the danger zone people would have no warning until it were too late.

For their own protection, the detection squads would enter the danger zone in motor-cars or aeroplanes which are armoured with lead plates, which absorb most of the dangerous radiation. The cabin would have to be hermetically sealed and oxygen carried in cylinders because of the danger from contaminated air.

The detection staff would have to know exactly the greatest dose of radiation to which a human being can be exposed safely for a short time. This safety limit is not at present known with sufficient accuracy and further biological research for this purpose is urgently required.

As regards the reliability of the conclusions outlined above, it may be said that they are not based on direct experiments, since nobody has ever yet built a super-bomb, but they are mostly based on facts which, by recent research in nuclear physics, have been very safely established. The only uncertainty concerns the critical size for the bomb. We are fairly confident that the critical size is roughly a pound or so, but for this estimate we have to rely on certain theoretical ideas which have not been positively confirmed. If the critical size were appreciably larger than we believe it to be, the technical difficulties in the way of constructing the bomb would be enhanced. The point can be definitely settled as soon as a small amount of uranium has been separated, and we think that in view of the importance of the matter immediate steps should be taken to reach at least this stage; meanwhile it is also possible to carry out certain experiments which, while they cannot settle the question with absolute finality, could, if their result were positive, give strong support to our conclusions.

Appendix III
Ralph Carlisle Smith's
Summary of the British
Mission at Los Alamos

(Unclassified with Deletions. Copy LANL Archives.)

July 18, 1949

To: Carroll L. Wilson
Through: Norris E. Bradbury
          Ralph Carlisle Smith
          British Mission
REF: LAB-ADCS-127

As requested, I have hurriedly collected information relative to the membership of and contributions by the British Mission (including Canadian) to the U.S. Atomic Energy Program while working at Los Alamos. While all the statements are my interpretation of the facts, they have been substantiated in substance by Hans Bethe, consultant, and formerly Theoretical Physics Division Leader at Los Alamos, and by Carson Mark, a former member of the British Mission and present leader of the Theoretical Physics Division of the Laboratory.

The contributions given are, of course, only the major ones which come to mind. They do not include the many which naturally result from senior staff members in the regular course of business. I believe the list of members actually working at Los Alamos is complete, but it may be checked from the monthly reports (May 1944 through March 1947) available to the Washington security staff which were written by the undersigned as Top Secret reports of the Commanding Officer to General Groves during the period of British participation in the program. These reports will give an indication of the extent of access to classified information. See especially the summary report of 18 September 1945 from Ralph Carlisle Smith to Thomas O. Jones.

The first British representatives to arrive at Los Alamos were Otto R. Frisch and Ernest Titterton (and wife), who joined the Project in December 1943. I believe the following summary is fairly complete.

Name	Initial Visit or Residence
Otto R. Frisch	13 December 1943
Ernest W. Titterton	13 December 1943
Mrs. E.W. Titterton	13 December 1943
Niels Bohr	30 December 1943

Aage Bohr	30	December 1943
Sir James Chadwick	12	January 1944
Rudolph Peierls	8	February 1944
Egon Bretscher	5	March 1944
M. Oliphant (no access)	19	March 1944
Philip B. Moon	25	March 1944
Mrs P.B. Moon	25	March 1944
W.L. Webster (no access)	25	March 1944
Joseph Rotblat	28	March 1944
Donald G. Marshall	15	April 1944
James L. Tuck	5	May 1944
Sir Geoffrey I. Taylor	24	May 1944
William J. Penney	29	June 1944
T.H.R. Skyrme	31	July 1944
Klaus Fuchs	14	August 1944
Lord Cherwell (Lindemann)	10	October 1944
W.G. Marley	14	October 1944
Michael J. Poole	17	October 1944
Anthony P. French	17	October 1944
James Hughes	17	October 1944
D. Littler	27	April 1945
H. Sheard	27	April 1945
George Placzek	7	May 1945
Carson Mark	29	May 1945
Boris Davison	22	October 1945
Lord Portal	18	May 1946
G.A. McMillan (no access)	8	June 1946

Permanent departures started about September 1945, although some early workers had left before this, e.g., Rotblat in December 1944. The last official representative was Ernest W. Titterton, who left the Laboratory in February 1947, Los Alamos in April 1947, and the country in August 1947. Of course George Placzek, who became an American citizen, and Carson Mark, who has indicated intention to become an American citizen and has his first papers, are still working with the Atomic Energy Commission.

Among the Project accomplishments of individuals of the British Mission are the following:

1. *Sir Geoffrey I. Taylor* worked with the Theoretical Division, and according to the leaders thereof, gave help, encouragement, leadership and advice in many phases of the work. For example, he gave great encouragement and suggestions in the work which led to a successful implosion. He knew where to find the data on high explosives, including information on the equations of state and on the theory of detonation waves. He introduced the workers into the field now known as Phenomenology, and personally predicted the phenomena which occurred at Trinity, including the mushroom, the heights reached, and the effect of wind on the distribution of particles. Many workers such as Penney were induced to come here by his efforts.
2. *Fuchs and Peierls* provided two-thirds of the team which handled the hydro-

dynamics in T Division which made the implosion development possible. They both contributed heavily to all phases of the weapon development, including implosion and Super.

3. *Fuchs* made efficiency estimates on various implosion designs, including composite structures – one of them corresponding rather closely to X-Ray shot at Eniwetok.

4. *Rudolph Peierls* was the joint inventor with Christy of the solid implosion gadget in which a modulated neutron source was used. Solid implosion had been suggested by von Neumann in his early patent application but the idea of a modulated neutron source with the solid implosion was that of *Peierls* and Christy, although we commonly call it the Christy gadget.

5. Also in the Theoretical Division were *George Placzek, Carson Mark, Boris Davisson, T.H.R. Skyrme*, all of whom contributed to the great success of that division.

6. *Tuck* independently and cooperatively with Neddermeyer and von Neumann suggested the lens system for symmetrical implosion.

7. *Tuck* independently suggested a modulated neutron source for the Christy gadget, employing the same principle of construction (but a different theory of operation) as that suggested by Bethe, resulting in the present Urchin. *Tuck's* theory of operation is presently favored, but the invention was made a joint one with Bethe.

8. *Tuck* worked on high velocity studies as a possible means of initiating Super.

9. *Tuck* and *Titterton* worked on the pin method for timing high explosive studies.

10. *Penney* contributed most in the field of use of the weapon, its effects, and their measurement. Matters relating to height of burst and conditions of burst were originally determined from his knowledge of bombings. He took part in the Survey Mission sent to Japan.

11. *Frisch* made many contributions, among which were the Dragon, early critical mass assembly studies, and water boiler design. Ingenious devices and methods of arranging fissionable material were suggested by Frisch. Independent of others, and the earliest record at Los Alamos, Frisch made the suggestion for adulterating the fissionable material with spontaneously fissionable isotopes to make it unsuitable for weapons. He also prepared long-range designs on nuclear reactors.

12. *Bretscher*, with the help of *French*, did considerable work on nuclear reactions of elements in a program of Super feasibility studies.

13. *Titterton* did outstanding work on electronic circuit developments relating to all fields of experimental work at Los Alamos, including high-voltage generators, detonator firing circuits, X-ray generator circuits, electric timing circuits, and coincidence circuits.

14. *Marley* brought his high-speed camera with him and contributed much in the field of high-speed photography and explosive lens development.

15. In the field of experimental nuclear physics, there were Moon, Marshall, Bretscher, Rotblat, Frisch, French, Hughes, Titterton, Tuck and Poole.

16. The advice of the senior scientists and Nobel prize winners, Niels Bohr and Sir James Chadwick, was priceless. Incidentally, Aage Bohr traveled as an aide to Niels Bohr.

17. Two Canadian representatives of the British Mission, George Placzek and Carson Mark, remained at Los Alamos as U.S. employees at the end of the war.

George Placzek was the first post-war division leader of the Theoretical Physics Division at Los Alamos. Carson Mark now occupies that important post.

18. All the British Mission personnel then at Los Alamos took part in the Trinity Operation either as observers or workers at or in preparation for the test. For example, *Littler* and *Sheard* made blast measurements, *Moon* worked on radiation measurements and *Titterton* was in charge of diagnostic electronic timing.

19. At the Bikini Operation, *Penney* was a technical advisor and made some blast measurements. *Tuck* made measurements on the time dependence of radiation and *Titterton* was in charge of technical electronic work.

20. *Hughes* assisted in the Fast Reactor development as a junior staff member.

From the access granted to British Mission personnel at Los Alamos, including working with U.S. employees formerly at other projects, plus the information gained by British Mission personnel at Columbia, Chicago, and Berkeley, the British (including Canadians) are well aware of all phases of the Manhattan Project, except possible details of Hanford and Oak Ridge construction. It is probably true that substantially none of the workers had much to do with the activities of the Chemistry and Metallurgy Divisions and the Fusing Group, but they were given access to substantially all local information.

The following information may help to identify some of the people. This is from recollection.

Sir James Chadwick, Nobel Prize winner for discovering the neutron. Professor of Physics at Liverpool University and now Master of Caius College, Cambridge University.

Niels Bohr, Nobel Prize winner. Identified with the Bohr-Wheeler Theory of Fission. Director of the Institute of Copenhagen. Substantially all the leading theoretical physicists have studied under Dr. Bohr.

Otto R. Frisch, with Lise Meitner, explained the results of Hahn and Strassmann as fission of the Uranium nucleus. First Head of Physics Division at Harwell Atomic Energy Research Establishment and left there to become Professor of Physics at Cambridge University.

Sir Geoffrey I. Taylor, Professor of Physics at Cambridge. High level Scientific Advisor to the British Government.

Rudolph Peierls, Professor of Physics at Birmingham University.

Klaus Fuchs, Head of Theoretical Group at Harwell, A.E.R.E.

William G. Penney, High level Scientific Advisor to the British Government in military matters.

James Tuck before coming here was assistant to Lindemann (Lord Cherwell), the scientific advisor to Churchill. Tuck presently is a Professor of Physics at Oxford.

Egon Bretscher now at Harwell A.E.R.E., formerly Cambridge University.

Ernest W. Titterton is now in charge of electronic work at Harwell A.E.R.E.

Philip B. Moon, Professor of Physics at Birmingham University.

Joseph Rotblat, equivalent to an Associate Professor at Liverpool University.

(signed)
RALPH CARLISLE SMITH

RCS:VJ

Appendix IV
Otto Frisch's Eyewitness Account of the July 16, 1945, Atomic Explosion at Trinity Site, Alamogordo Air Base, New Mexico

[Original in Foreign Relations of the United States; *Potsdam* (Washington: USGPO, 1960), 1371.]

I watched the explosion from a point said to be about 20 (or 25) miles away and about north of it, together with the members of the co-ordinating council. Fearing to be dazzled and to be burned by ultraviolet rays, I stood with my back to the gadget, and behind the radio truck. I looked at the hills, which were visible in the first faint light of dawn (0530 M.W. Time). Suddenly and without any sound, the hills were bathed in brilliant light, as if somebody had turned the sun on with a switch. It is hard to say whether the light was less or more brilliant than full sunlight, since my eyes were pretty well dark adapted. The hills appeared kind of flat and colourless like a scenery seen by the light of a photographic flash, indicating presumably that the retina was stimulated beyond the point where intensity discrimination is adequate. The light appeared to remain constant for about one or two seconds (probably for the same reason) and then began to diminish rapidly.

After that I turned round and tried to look at the light source but found it still too bright to keep my eyes on it. A few short glances gave me the impression of a small very brilliant core much smaller in appearance than the sun, surrounded by decreasing and reddening brightness with no definite boundary, but not greater than the sun. After some seconds I could keep my eye on the thing and it now looked like a pretty perfect red ball, about as big as the sun, and connected to the ground by a short grey stem. The ball rose slowly, lengthening its stem and getting gradually darker and slightly larger. A structure of darker and lighter irregularities became visible, making the ball look somewhat like a raspberry. Then its motion slowed down and it flattened out, but still remained connected to the ground by its stem, looking more than ever like the trunk of an elephant. Then a hump grew out of its top surface and a second mushroom grew out of the top of the first one, slowly penetrating the highest cloud layers. As the red glow died out it became apparent that the whole structure, in particular the top mushroom, was surrounded by a purplish blue glow. A minute or so later the whole top mushroom appeared to glow feebly in this colour, but this was no longer easy to see, in the increasing light of dawn.

A very striking phenomenon was the sudden appearance of a white patch on the underside of the cloud layer just above the explosion; the patch spread very rapidly, like a pool of spilt milk, and a second or two later, a similar patch appeared and spread on another cloud layer higher up. They marked no doubt the impact of the blast wave on the cloud layers. They appeared, I believe, before the red ball had started to flatten out.

When I thought it was soon time for the blast to arrive, I sat on the ground, still facing the explosion, and put my fingers in my ears. Despite that, the report was quite respectable and was followed by a long rumbling, not quite like thunder but more regular, like huge noisy wagons running around in the hills.

Bibliography

PRIMARY SOURCES

A. The Archives of the Los Alamos National Laboratory, Los Alamos, New Mexico.
B. The papers of Klaus Fuchs (65,570 pages) released by the FBI.
C. Materials at the Harry S Truman Library, Independence, Missouri.
D. Materials at the Public Record Office, Kew, London.
E. Atomic Bomb folder, from the Franklin D. Roosevelt Library, Hyde Park, New York.
F. The Ralph Carlisle Smith materials, Special Collections, Zimmerman Library, The University of New Mexico.
G. Unpublished essay: Albrecht, Ulrich, "The Development of the First Atomic Bomb in the USSR," Paper for the Conference at Harvard University, Cambridge, January 8–10, 1987.

ARTICLES

Anders, Roger M., "The President and the Atomic Bomb: Who Approved the Trinity Nuclear Test?" *Prologue*, Vol. 20 (Winter 1988), pp. 268–82.
Almaraz, Felix Diaz, Jr., "The Little Theatre in the Atomic Age: Amateur Dramatics in Los Alamos, New Mexico, 1943–1946," *Journal of the West*, Vol. 17 (1978), pp. 72–8.
Bernstein, Barton J., "The Quest for Security: American Foreign Policy and International Control of Atomic Energy, 1942–1946," *Journal of American History*, Vol. LX (March 1974), pp. 1003–44.
———, "The Uneasy Alliance: Roosevelt, Churchill, and the Atomic Bomb, 1940–1945," *Western Political Quarterly*, Vol. 29 (1976), pp. 202–30.
Bethe, Hans, "Chop Down the Nuclear Arsenals," *Bulletin of the Atomic Scientists*, Vol. 45 (March 1989), pp. 11–15.
Breindel, Eric M., "Do Spies Matter?" *Commentary*, Vol. 85 (March 1988), pp. 53–8.
"British Developments in Atomic Energy," *Nucleonics*, Vol. 12 (January 1954), pp. 6–10.
Bush, Vannevar, "Churchill and the Scientists," *Atlantic Monthly*, Vol. CCXU (March 1965), pp. 94–100.
"Did the Soviet Bomb Come Sooner than Expected?" *Bulletin of the Atomic Scientists*, Vol. V (October 1949), pp. 263–73.
Duncan, Francis, "Atomic Energy and Anglo American Relations, 1946–1954," *Orbis*, Vol. 12 (Fall 1968), pp. 1188–1201.
Frisch, Otto R., "How It All Began," *Physics Today*, Vol. 20 (November 1967), pp. 43–8.

"Fuchs at Los Alamos," *The Spectator*, Vol. CCIII (September 18, 1959), pp. 363–6.

Gowing, Margaret, "Niels Bohr and Nuclear Weapons," in A.P. French and P.J. Kennedy (eds), *Niels Bohr: A Centenary Volume*. Cambridge, Mass.: Harvard University Press, 1985, pp. 266–77.

———, "Reflections on Atomic Energy History," *Bulletin of the Atomic Scientists*, Vol. 35 (March 1979), pp. 51–4.

Goldberg, Alfred, "The Atomic Origins of the British Nuclear Deterrent," *International Affairs*, Vol. 40 (July 1964), pp. 409–29.

Gormly, James L., "The Washington Declaration and the 'Poor Relation': Anglo-American Atomic Diplomacy, 1945–46," *Diplomatic History*, Vol. 8 (Spring 1984), pp. 125–43.

Graybar, Lloyd J., "The 1946 Atomic Bomb Tests: Atomic Diplomacy or Bureaucratic Infighting?" *Journal of American History*, Vol. 72 (March 1986), pp. 888–907.

Hahn, Otto, "The Discovery of Fission," *Scientific American*, Vol. 198 (February 1958), pp. 76–84.

Herken, Gregg, "'A Most Deadly Illusion': The Atomic Secret and American Nuclear Weapons Policy, 1945–1950," *Pacific Historical Review*, Vol. XLIX (February 1980), pp. 51–76.

Hirsch, Daniel and William G. Mathews, "The H-Bomb: Who Really Gave Away the Secret?" *Bulletin of the Atomic Scientists*, Vol. 46 (January–February, 1990), pp. 22–30.

Holloway, David, "Entering the Nuclear Arms Race: The Soviet Decision to Build the Atomic Bomb, 1939–1945," *Social Studies of Science*, Vol 11 (1981), pp. 159–97.

Laurence, George C., "Canada's Participation in Atomic Energy Development," *Bulletin of the Atomic Scientists*, Vol. 3 (November 1947), pp. 325–9.

Manley, J.H., "Assembling the Wartime Labs," *Bulletin of the Atomic Scientists*, Vol. 30 (1974), pp. 42–7.

Mark, Eduard, "'Today Has Been a Historical One': Harry S Truman's Diary of the Potsdam Conference," *Diplomatic History*, Vol. 4 (1980), pp. 317–26.

Meitner, Lise, "Looking Back," *Bulletin of the Atomic Scientists* (November 1964), pp. 2–6.

———, "Right and Wrong Roads to the Discovery of Nuclear Energy," *International Atomic Energy Bulletin*, (December 2, 1962), p. 608.

Miller, Byron S., "A Law Is Passed – The Atomic Energy Act of 1946," *The University of Chicago Law Review*, Vol. 15 (Summer 1948), pp. 799–821.

Oppenheimer, Robert, "Niels Bohr and Atomic Weapons," *New York Review of Books*, Vol. III (December 17, 1964), pp. 6–8.

Peierls, Sir Rudolph, "Britain in the Atomic Age," in Richard S. Lewis and Jane Wilson (eds), *Alamogordo Plus Twenty-five Years*. New York: Viking Press, 1970.

Rosenberg, David Alan, "A Smoking Radiating Ruin at the End Two Hours, Documents on American Plans for Nuclear War with the Soviet Union, 1954–1955," *International Security*, Vol. 6 (Winter 1981–2), pp. 1–37.

———, "U.S. Nuclear Stockpile, 1945 to 1950," *Bulletin of the Atomic Scientists*, Vol. 38 (May 1982), pp. 25–30.

Rotblat, Joseph. "British Fret about 'Vulnerability,'" *Bulletin of the Atomic Scientists*, Vol. 44 (March 1988), pp. 20–2.

Smith, Alice Kimball, "Los Alamos: Focus of an Age," *Bulletin of the Atomic Scientists*, Vol. 26 (June 1970), pp. 15–20.

Stuewer, Roger H., "Bringing the News of Fission to America," *Physics Today*, Vol. 38 (October 1958), pp. 49–55.

Villa, B.L., "The Atomic Bomb and the Normandy Invasion," *Perspectives in American History*, Vol. 11 (1977–8), pp. 463–502.

Weart, Spencer R., "Scientists in Power: France and the Origins of Nuclear Energy, 1900–1950," *Bulletin of the Atomic Scientists* (March 1979), pp. 41–50.

Zuckerman, Lord, "Nuclear Wizards," *New York Review of Books*, March 13, 1988, pp. 26–31.

BOOKS

Ackland, Len and Steven McGuire (eds), *Assessing the Nuclear Age*. Chicago: Educational Foundation for Nuclear Science, 1986.

Anders, Roger M. (ed.), *Forging the Atomic Shield: Excerpts from the Office Diary of Gordon E. Dean*. Chapel Hill: University of North Carolina Press, 1967.

Alvarez, Luis W., *Alvarez: Adventures of a Physicist*. New York: Basic Books, 1987.

Amrine, Michael, *The Great Decision: The Secret History of the Atomic Bomb*. New York: G.P. Putnam's, 1959.

Arnold, Lorna, *A Very Special Relationship: British Atomic Weapons Trials in Australia*. London: HMSO, 1987.

Atomic Challenge: A Symposium. London: Winchester Publications, 1946.

Badash, Lawrence, *Kapitsa, Rutherford and the Kremlin*. New Haven: Yale University Press, 1985.

——— et al. (eds), *Reminiscences of Los Alamos, 1943–1945*. Dordrecht, Holland: D. Reidel, 1980.

Baxter, James Phinney III, *Scientists Against Time*. Cambridge: MIT Press, 1968; 1946.

Baylis, John, *Anglo-American Defense Relations, 1939–1984: The Special Relationship*. New York: St. Martin's Press, 1984.

Bernstein, Jeremy, *Hans Bethe: Prophet of Energy*. New York: Basic Books, 1979.

Birkenhead, The Earl of, *The Prof. in Two Worlds: The Official Life of Professor F.A. Lindemann, Viscount Cherwell*. London: Collins, 1961.

Blaedel, Niels, *Harmony and Unity: The Life of Niels Bohr*, trans. Geoffrey French. Madison, Wis.: Science Tech, 1988.

Bohr, Niels, *Atomic Physics and Human Knowledge*. New York: John Wiley, 1958.

———, *The Philosophical Writings of Niels Bohr*, Vol. II: *Essays 1932–1957 on Atomic Physics and Human Knowledge*. Woodbridge, Conn.: Ox Bow Press, 1987.

———, *The Philosophical Writings of Niels Bohr*, Vol. III: *Essays 1958–1962 on Atomic Physics and Human Knowledge*. Woodbridge, Conn.: Ox Bow Press, 1963.

Born, Max, *My Life: Recollections of a Nobel Laureate*. London: Taylor & Francis, 1978.

Botti, Timothy J., *The Long Wait: The Forging of the Anglo-American Nuclear Alliance, 1945–1958*. New York: Greenwood, 1987.

Boyer, Paul, *By the Bomb's Early Light: American Thought and Culture at the*

Dawn of the Atomic Age. New York: Pantheon, 1985.

Brinkley, David, *Washington Goes to War.* New York: Alfred A. Knopf, 1988.

Bundy, McGeorge, *Danger and Survival: Choices about the Bomb in the First Fifty Years.* New York: Random House, 1988.

Calder, Angus, *The People's War: Britain, 1939–45.* London: Panther, 1971.

Caldicott, Helen, *Missile Envy: The Arms Race and Nuclear War.* New York: William Morrow, 1984.

Caute, David, *The Great Fear: The Anti-Communist Purge Under Truman and Eisenhower.* New York: Simon & Schuster, 1978.

Churchill, Winston S., *The Hinge of Fate.* Boston: Houghton Mifflin, 1950.

———, *The Gathering Storm.* Boston: Houghton Mifflin, 1948.

———, *The Grand Alliance.* Boston: Houghton Mifflin, 1950.

———, *Triumph and Tragedy.* Boston: Houghton Mifflin, 1953.

Clarfield, Gerald H. and William M. Wiecek, *Nuclear America: Military and Civilian Nuclear Power in the United States, 1940–1980.* New York: Harper & Row, 1984.

Clark, Ronald W., *The Birth of the Bomb: The Untold Story of Britain's Part in the Weapon That Changed the World.* London: Phoenix House, 1961.

———, *The Greatest Power on Earth: The International Race for Nuclear Supremacy.* New York: Harper & Row, 1980.

Cockburn, Stewart, and David Ellyard, *Oliphant: The Life and Times of Sir Mark Oliphant.* Adelaide: Axiom Books, 1981.

Costello, John, *Mask of Treachery.* New York: Morrow, 1988.

Dallek, Robert. *Franklin D. Roosevelt and American Foreign Policy, 1932–1945.* New York: Oxford University Press, 1979.

Danchev, Alex, *Very Special Relationship: Field-Marshal Sir John Dill and the Anglo-American Alliance, 1941–44.* London: Brassey's Defence Publishers, 1986.

Deacon, Richard, *A History of the British Secret Service.* New York: Taplinger, 1969.

Del Tredici, Robert. *At Work in the Fields of the Bomb.* New York: Harper & Row, 1987.

Dilks, David (ed.), *Retreat from Power: Studies in Britain's Foreign Policy of the Twentieth Century*, Vol. II: *After 1939.* London: Macmillan, 1981.

Divine, Robert A., *Blowing on the Wind: The Nuclear Test Ban Debate, 1954–1960.* New York: Oxford University Press, 1978.

Donovan, Robert J., *Conflict and Crisis: The Presidency of Harry S Truman, 1945–1952.* New York: W.W. Norton, 1977.

Feis, Herbert, *The Atomic Bomb and the End of World War II.* Princeton: Princeton University Press, 1961; 1966.

Fermi, Laura, *Atoms in the Family: My Life with Enrico Fermi.* Chicago: University of Chicago Press, 1954.

Feynman, Richard P., *"What Do You Care What Other People Think?" Further Adventures of a Curious Character.* New York: W.W. Norton, 1988.

Fisher, David E., *A Race on the Edge of Time: Radar – The Decisive Weapon of World War II.* New York: McGraw-Hill, 1988.

Fisher, Nigel, *Harold Macmillan.* London: Weidenfeld & Nicolson, 1982.

Fisher, Phyllis, *Los Alamos Experience.* Tokyo and New York: Japan Publications, 1985.

Foreign Relations of the United States: Conference at Washington, 1941–1942, and Casablanca, 1943. Washington: USGPO, 1968.

Freedman, Lawrence, *The Evolution of Nuclear Strategy*. New York: St. Martin's Press, 1981.

French, A.P. and P.J. Kennedy (eds), *Niels Bohr: A Centenary Volume*. Cambridge, Mass.: Harvard University Press, 1985.

Frisch, Otto R. *What Little I Remember*. Cambridge: Cambridge University Press, 1979.

Furman, Necah Stewart, *A History of Sandia National Laboratories – The Postwar Decade*. Albuquerque: University of New Mexico Press, 1990.

Fussell, Paul, *Thank God for the Atomic Bomb and Other Essays*. New York: Summit Books, 1988.

Gardner, Brian, *The Year That Changed the World, 1945*. New York: Coward-McCann, 1963.

Gilbert, Martin, *"Never Despair," Winston S. Churchill, 1945–1965*. London: Heinemann, 1988.

Gilbert, Martin, *Road to Victory, 1941–1945*. Boston: Houghton Mifflin, 1986.

Goodchild, Peter., *J. Robert Oppenheimer, Shatterer of Worlds*. Boston: Houghton Mifflin, 1981.

Goldschmidt, Bertrand, *The Atomic Adventure*, trans. from the French by Peter Beer. Oxford: Pergamon, 1964.

Goudsmit, Samuel A., *Alsos*. New York: Henry Schuman, 1947.

Gowing, Margaret, *Britain and Atomic Energy, 1939–1945*. London: Macmillan, 1964.

———, *Independence and Deterrence: Britain and Atomic Energy, 1945–1953*, Vol. I: *Policy Making*; Vol. II: *Policy Execution*. London: Macmillan, 1974.

Grinspoon, Lester (ed.), *The Long Darkness: Psychological and Moral Perspectives on Nuclear Winter*. New Haven: Yale University Press, 1986.

Grodzins, Morton, and Eugene Rabinowitch (eds), *The Atomic Age: Scientists in National and World Affairs*. New York: Basic Books, 1963.

Groom, A.J.R., *British Thinking About Nuclear Weapons*. London: Francis Pinter, 1974.

Groueff, Stephane. *Manhattan Project: The Untold Story of the Making of the Atomic Bomb*. New York: Bantam, 1967.

Groves, Leslie R., *Now It Can Be Told: The Story of the Manhattan Project*. New York: Da Capo Press, 1962; 1983.

Hacker, Barton C., *The Dragon's Tail: Radiation Safety in the Manhattan Project, 1942–1946*. Berkeley: University of California Press, 1987.

Hansen, Chuck, *U.S. Nuclear Weapons: The Secret History*. New York: Crown, 1987.

Hathaway, Robert M., *Ambiguous Partnership: Britain and America, 1944–1947*. New York: Columbia University Press, 1981.

Hawkins, David, *Toward Trinity*, Part 1 of *Project Y: The Los Alamos Story*. Los Angeles: E.A. Tomash, 1983.

Hennessey, Peter, *Whitehall*. London: Secker & Warburg, 1989.

Herken, Gregg, *Counsels of War*. New York: Alfred A. Knopf, 1985.

———, *The Winning Weapon: The Atomic Bomb in the Cold War, 1945–1950*. New York: Alfred A. Knopf, 1988.

Hewlett, Richard G. and Oscar E. Anderson, Jr., *A History of the United States Atomic Energy Commission*, Vol. I: *The New World, 1939/1946*. University Park, Penn.: Pennsylvania State University Press, 1962.

Hewlett, Richard G. and Francis Duncan, *Atomic Shield, 1947/1952*, Vol. II of *A*

History of the United States Atomic Energy Commission. University Park, Penn., and London: Pennsylvania State University Press, 1969.

―――― and Jack Holl. *Atoms for Peace and War, 1953–1961: Eisenhower and the Atomic Energy Commission.* Berkeley: University of California Press, 1989.

Hinsley, Francis Harry, *British Intelligence in the Second World War: Its Influence on Strategy and Operations,* Vol. I. London: HMSO, 1979.

Holloway, David, *The Soviet Union and the Arms Race.* New Haven, Conn.: Yale University Press, 1983; 1986.

Hoover, J. Edgar, *Masters of Deceit.* New York: Henry Holt, 1958.

Horne, Alistair, *Harold Macmillan,* Vol. II: *1957–1986.* New York: Viking, 1989.

Hyde, H. Montgomery, *The Atom Bomb Spies.* London: Hamish Hamilton, 1980.

In the Matter of J. Robert Oppenheimer: Transcript of Hearing Before Personnel Security Board and Texts of Principal Documents and Letters. Cambridge, Mass.: MIT Press, 1971.

Jackman, Jarrell C. and Carla M. Borden (eds), *The Muses Flee Hitler: Cultural Transfer and Adaptation, 1930–1945.* Washington: Smithsonian Institution Press, 1983.

Jones, R.V., *The Wizard War: British Scientific Intelligence, 1939–1945.* New York: Coward, McCann & Geoghegan, 1978.

Jones, Vincent C., *Manhattan: The Army and the Atomic Bomb.* Washington: USGPO, 1984.

Jungk, Robert. *Brighter Than a Thousand Suns: A Personal History of the Atomic Scientists.* New York: Harcourt, Brace, 1956.

Kegley, Charles W., Jr. and Eugene R. Wittkopf (eds), *The Nuclear Reader: Strategy, Weapons, War.* New York: St. Martin's Press, 1985.

Kennett, Lee, *A History of Strategic Bombing.* New York: Scribner's, 1982.

Kevles, Daniel J., *The Physicists: the History of a Scientific Community in Modern America.* New York: Alfred A. Knopf, 1977.

Kimball, Warren F. (ed.), *Churchill and Roosevelt: The Complete Correspondence,* 3 volumes. Princeton: Princeton University Press, 1984.

Knightley, Philip, *The Second Oldest Profession: Spies and Spying in the Twentieth Century.* New York: W.W. Norton, 1986.

Kramish, Arnold, *The Griffin: The Greatest Untold Espionage Story of World War II.* Boston: Houghton Mifflin, 1986.

Kunetka, James W., *City of Fire: Los Alamos and the Birth of the Atomic Age, 1943–1945.* Albuquerque: University of New Mexico Press, 1979.

Kunetka, James W., *Oppenheimer: The Years of Risk.* Englewood Cliffs, NJ: Prentice Hall, 1982.

Kurzman, Dan, *Day of the Bomb: Countdown to Hiroshima.* New York: McGraw-Hill, 1986.

Lamont, Lansing, *Day of Trinity.* New York: Atheneum, 1965.

Lamphere, Robert J. and Tom Shachtman, *The FBI–KGB War: A Special Agent's Story.* New York: Random House, 1986.

Lawren, William, *The General and the Bomb: A Biography of General Leslie R. Groves, Director of the Manhattan Project.* New York: Dodd, Mead, 1988.

Lewis, Richard S. and Jane Wilson (eds), *Alamogordo Plus Twenty-Five Years.* New York: Viking, 1970.

Lilienthal, David E., *The Journals of David E. Lilienthal,* Vol. II: *The Atomic Energy Years, 1945–1950.* New York: Harper & Row, 1964.

Louis, William Roger and Hedley Bull (eds), *The 'Special Relationship': Anglo-American Relations Since 1945*. New York: Oxford, 1986.

McArdle, Catherine *et al.* (eds), *Nuclear Deterrence: New Risks. New Opportunities*. Washington: Pergamon-Brassey's, 1986.

Macmillan, Harold, *Riding the Storm, 1956–1959*. New York: Harper & Row, 1971.

McNeill, William H., *America, Britain, and Russia: Their Co-operation and Conflict*. New York: Oxford, 1953.

Malone, Peter, *The British Nuclear Deterrent*. New York: St. Martin's Press, 1984.

Milliken, Robert, *No Conceivable Injury: The Story of Britain and Australia's Atomic Cover-Up*. Ringwood, Victoria, Australia: Penguin, 1986.

Moore, Ruth, *Niels Bohr: The Man, His Science, and the World They Changed*. Cambridge, Mass.: MIT Press, 1985.

Moss, Norman, *Klaus Fuchs: A Biography*. New York: St. Martin's Press, 1987.

———, *The Politics of Uranium*. New York: Universe Books, 1982.

Nichols, H.G. (ed.), *Washington Dispatches, 1941–1945: Weekly Political Reports from the British Embassy*. Chicago: University of Chicago Press, 1981.

Nichols, K.D., *The Road to Trinity: A Personal Account of How America's Nuclear Policies Were Made*. New York: William Morrow, 1987.

Nissen, Jack and A.W. Cockerill, *Winning the Radar War, 1939–1945*. New York: St. Martin's Press, 1987.

Nuclear America: A Historical Bibliography. Santa Barbara, Calif.: ABC Information Services, 1984.

One World or None New York: McGraw-Hill, 1946.

Pash, Boris T., *The Alsos Mission*. New York: Award House, 1965.

Pawle, Gerald, *The War and Colonel Warden*. London: Harrap, 1963.

Peierls, Rudolf, *Bird of Passage: Recollections of a Physicist*. Princeton: Princeton University Press, 1985.

Pierre, Andrew J., *Nuclear Politics: The British Experience with an Independent Strategic Force, 1939–1970*. Oxford University Press, 1972.

Pickersgill, J.W., *The Mackenzie King Record, 1939–1944*. Vol. I of IV. Toronto: University of Toronto Press and University of Chicago Press, 1960.

Pilpel, Robert H., *Churchill in America, 1895–1961: An Affectionate Portrait*. New York: Harcourt, Brace, Jovanovich, 1976.

Pilat, Oliver, *The Atom Spies*. New York: G.P. Putnam's, 1952.

Pincher, Chapman, *Too Secret Too Long*. New York: St. Martin's Press, 1984.

———, *Traitors: The Anatomy of Treason*. New York: St. Martin's Press, 1987.

Philby, Kim, *My Silent War*. New York: Ballantine, 1968.

Powaski, Ronald E., *March to Armageddon: The United States and the Nuclear Arms Race, 1939 to the Present*. New York: Oxford, 1987.

Powers, Richard Gid, *Secrecy and Power: The Life of J. Edgar Hoover*. New York: Free Press, 1987.

Psychosocial Aspects of Nuclear Developments: A Report of the Task Force on Psychosocial Aspects of Nuclear Developments of the American Psychiatric Association. Washington: American Psychiatric Association, 1982.

Radosh, Ronald and Joyce Milton, *The Rosenberg File: A Search for Truth*. New York: Vintage, 1984.

Reid, R.W., *Tongues of Conscience: War and the Scientists' Dilemma*. London: Readers Union, 1970.

Rhodes, Richard, *The Making of the Atomic Bomb*. New York: Simon & Schuster, 1986.

Rigden, John S., *Rabi: Scientist and Citizen*. New York: Basic Books, 1987.

Rotblat, Joseph, *History of the Pugwash Conferences*. London: Taylor & Francis, 1962.

Rozental, S. (ed.), *Niels Bohr*. Amsterdam: North-Holland, 1967.

Rupke, Nicolaas A. (ed.), *Science, Politics and the Public Good: Essays in Honour of Margaret Gowing*. London: Macmillan, 1988.

Ryan, Henry Butterfield, *The Vision of Anglo-America: The US–UK Alliance and the Emerging Cold War, 1943–1946*. Cambridge: Cambridge University Press, 1987.

Sherry, Michael S., *The Rise of American Air Power: The Creation of Armageddon*. New Haven, Conn.: Yale University Press, 1987.

Sherwin, Martin, *A World Destroyed: The Atomic Bomb and the Grand Alliance*. New York: Knopf, 1975.

Shurcliff, W.A., *Bombs at Bikini: The Official Report of Operation Crossroads*. New York: Wm. H. Wise, 1947.

Simpson, John, *The Independent Nuclear State: The United States, Britain and the Military Atom*. New York: St. Martin's Press, 1983.

Smith, Alice Kimball, *A Peril and a Hope: The Scientists' Movement in America: 1945–47*. Cambridge, Mass.: MIT Press, 1965, 1970.

―――― and Charles Weiner (eds), *Robert Oppenheimer: Letters and Recollections*. Cambridge and London: Harvard University Press, 1980.

Smyth, Henry DeWolf, *Atomic Energy for Military Purposes: The Official Report on the Development of the Atomic Bomb under the Auspices of the United States Government, 1940–1945*. Princeton: Princeton University Press, 1945.

Statements Relating to the Atomic Bomb. London: HMSO, 1945.

Stuewer, Roger H. (ed.), *Nuclear Physics in Retrospect: Proceedings of a Symposium on the 1930s*. Minneapolis: University of Minnesota Press, 1979.

Sutherland, Douglas, *The Great Betrayal: The Definitive Story of the Most Sensational Spy Case of the Century*. New York: Penguin, 1980.

Symonds, J.L., *A History of British Atomic Tests in Australia*. Canberra: Australian Government Publishing Service, 1985.

Szasz, Ferenc Morton, *The Day the Sun Rose Twice: The Story of the Trinity Site Nuclear Explosion, July 16, 1945*. Albuquerque: University of New Mexico Press, 1984.

Teller, Edward, *Better a Shield Than a Sword: Perspectives on Defense and Technology*. New York: Macmillan, 1987.

―――― with Allen Brown, *The Legacy of Hiroshima*. Garden City, NY: Doubleday, 1962.

Thompson, James, *Psychological Aspects of Nuclear War*. New York: The British Psychological Society and John Wiley & Sons, 1985.

Trower, W. Peter (ed.), *Discovering Alvarez: Selected Works of Luis W. Alvarez with Commentary by His Students and Colleagues*. Chicago: University of Chicago Press, 1987.

Truslow, Edith C. and Ralph Carlisle Smith, *Beyond Trinity*, Part II of *Project Y: The Los Alamos Story*. Los Angeles: Tomash, 1983.

Walker, Mark, *German National Socialism and the Quest for Nuclear Power*. New York: Cambridge University Press, 1989.

Wasserstein, Bernard, *Britain and the Jews of Europe, 1939–1945*. Oxford: Clarendon Press, 1979.

Watt, Donald Cameron, *How War Came: The Immediate Origins of the Second World War, 1938–1939*. New York: Pantheon, 1989.

Weart, Spencer R., *Nuclear Fear: A History of Images*. Cambridge: Harvard University Press, 1988.

—————— and Gertrud Weiss Szilard (eds), *Leo Szilard: His Version of the Facts*. Cambridge, Mass.: MIT Press, 1978.

Weisskopf, Victor F., *The Privilege of Being a Physicist*. New York: W.H. Freeman, 1989.

West, Nigel, *The Circus: MI5 Operations, 1945–1972*. New York: Stein & Day, 1983.

——————, *MI6: British Secret Intelligence Service Operations, 1909–45*. New York: Random House, 1983.

West, Rebecca, *The New Meaning of Treason*. New York: Viking, 1964; 1985.

Wheeler-Bennett, John W., *John Anderson, Viscount Waverley*. New York: St. Martin's Press, 1962.

Williams, Robert Chadwell, *Klaus Fuchs, Atom Spy*. Cambridge and London: Harvard University Press, 1987.

Williams, Robert C. and Philip L. Cantelon (eds), *The American Atom: A Documentary History of Nuclear Policies from the Discovery of Fission to the Present, 1939–1984*. Philadelphia: University of Pennsylvania Press, 1984.

Williamson, Rajkumari (ed.), *The Making of Physicists*. Bristol: Adam Hilger, 1987.

Wilson, Jane S. and Charlotte Serber (eds), *Standing By and Making Do: Women of Wartime Los Alamos*. Los Alamos, NM: The Los Alamos Historical Society, 1988.

Woods, Randall Bennett, *A Changing of the Guard: Anglo-American Relations, 1941–1946*. Chapel Hill, NC: University of North Carolina Press, 1990.

Wright, Peter, *Spycatcher: The Candid Autobiography of a Senior Intelligence Officer*. New York: Viking, 1987.

Wyden, Peter, *Day One: Before Hiroshima and After*. New York: Simon & Schuster, 1984.

York, Herbert, *The Advisors: Oppenheimer, Teller and the Superbomb*. San Francisco: W.H. Freeman, 1976.

——————, *Making Weapons, Talking Peace: A Physicists's Odyssey from Hiroshima to Geneva*. New York: Basic Books, 1987.

Index

163